LEARNING RESOURCES CTR/NEW ENGLAND TECH.
GEN HD8039.A8 F67
Form, Willia Blue-collar stratification
3 0147 0000 5048 7

W9-ABK-874

Blue-Collar Stratification

Blue-Collar Stratification

Autoworkers in Four Countries

WILLIAM H. FORM

PRINCETON UNIVERSITY PRESS · 1976

Published by Princeton University Press, Princeton, New Jersey
In the United Kingdom: Princeton University Press, Guildford, Surrey

Library of Congress Cataloging in Publication Data will
be found on the last printed page of this book
This book has been composed in Linotype Caledonia
Printed in the United States of America
by Princeton University Press, Princeton, New Jersey

To my brother and teacher

George H. Form

Contents

List of Tables

Preface

After World War II universities followed a fashion of combining departments of sociology, social psychology, and anthropology, whose faculties were obliged to know all three of these interrelated fields. In such a department I was asked to teach social psychology, a subject I knew little about. In good spirit, I taught courses for five years, but became increasingly dissatisfied with the endless studies of attitudes, motivation, and identification which seemed unrelated to what was going on in the world. Anthropologists in the department, who were in ascendancy, constantly talked about culturological explanations. At first I listened with deference, but later decided that when I liked what they said they were being sociologists. My first comparative study of the U.S.-Mexican border convinced me that technology and stratification account for almost everything and that most social psychological and culturological explanations bite the dust when technology and stratification are taken into account. The way to demonstrate this was to study a single industry in different countries. This is how the present research was born and how, after a number of years, I became a chastened, cautious, apologetic sociologist. The world turned out to be more complicated than I had imagined.

The appearance of Alex Inkeles' path-breaking article, "Industrial Man," encouraged me to act. I was irritated that he had beaten me to demonstrating that workers in different countries hold similar attitudes and beliefs, but I was encouraged to believe that a cross-national study which controlled technology more precisely than Inkeles did would provide even more dramatic evidence of the power of a technological-structural perspective. I was acquainted with the technology of the automobile industry (Oldsmobile) in Lansing, Michigan and had studied the same factory made famous by Chinoy's *Automobile Workers and the American Dream.* I decided to take a sabbatical in Turin to study FIAT and to compare it to Oldsmobile.

A direct comparison of American and Italian automobile workers would provide a good test of the structural perspective. Italy was less industrial than the United States, Italians were more exposed to ideological unionism, Italy was a more highly stratified society than the United States, Italy had a stronger artisan tradition, and its socio-political system produced more cynics. I was convinced I would find few differences between the Italian and American workers. When people asked me what I was going to study, I replied, "Italian automobile workers and the American Dream." But I was really more interested in a broader question involving urban sociology and social stratification. The big question in urban sociology had been, "How do people who presumably live in the pathological urban environment manage to live such organized lives?" To answer this, I would have to study how workers live in various social systems: family, work place, neighborhood, union, community, and nation. Stratification theory suggested that system involvements would differ according to income and skill even among manual workers. Thus the main question of the research would be, "How do skill differences within the factory affect the social system involvements of workers in the same industry in different societies?"

The Italian experience was sobering. The FIAT works in Turin were impressively modern; FIAT employees were mostly northerners who were perhaps as much exposed to industrial culture as most Americans; Turin was a cosmopolitan city which made Lansing look like a provincial town, and multi-unionism in Italy was more Byzantine than I had imagined. Though preliminary analysis of the data seemed to support the theoretical orientation of the primacy of technological variables, I felt uneasy and looked for an opportunity to expand the study with other cases in less developed societies. I looked to Latin America, where my earlier research experiences had convinced me that almost any automobile factory there would have workers less caught up in the modern industrial culture and where urban facilities would be more primitive than in Lansing or Turin. I found a factory in Córdoba, Argentina, which manufactured automobiles and trucks. A visit to the city and the factory convinced me that a strict adherence to the original research design was impossible. Though the factory was run almost exclusively by Argentinians, its technology was essentially an earlier American version of a Rambler-Jeep factory: I could not rigidly hold technology constant and have the other variations that I wanted. Yet the factory did produce automobiles, and workers in certain departments were performing tasks very similar to those of workers in the

United States and Italy. But Argentinians were more urbanized than respondents in the two other countries. They were better educated than the Italians, and the city was more cosmopolitan than Lansing.

The growing difficulty in controlling the technological variable was explicitly faced in the Indian case. Here the machinery of an old FIAT factory was still being used long after it would have been discarded, even in Latin America. But for Indian manufacturing, the plant represented a relatively modern exemplar and cars and trucks were being produced by workers with skills similar to those of workers in the other countries. Clearly, job routines could be quite similar within factories which had equipment which ranged widely in modernity. Importantly, the Indian case gave the study needed non-factory variables: the Indian workers had not been socialized in an industrial environment; they were quite heterogeneous in their religious, caste, community, and regional backgrounds; and they had little or no conception of what social science research was about.

The sober lesson learned was that it was almost impossible to control technology in societies which are so different in extent of industrialization. Since the major control variable turned out to be quite mushy, the industrialism hypothesis was too crude. In the attempt to make a virtue out of the inevitable, I turned to the development literature. The technological differences within the four factories were certainly smaller than the differences in extent of modern technology in the larger societies. The task then became one of accounting for the differences in responses of workers to small differences in the technologies of the factories. Even though I expected response differences to be small, I also expected them to vary systematically along the dimension of technological complexity. This made the research much more exciting, but reporting more difficult. Technology now had three dimensions: extent of prevalence in the societies, range of complexity among the factories, and range of complexity of work operations within each factory.

Little is gained by previewing the findings, but it may be useful to list some hard-learned lessons. One purpose of the Preface is to tell readers about the monumental difficulties I confronted in the research, in the hope of being forgiven for errors. But the ruse rarely works; criticism is built into academia. But I do have some unsolicited advice for people who want to do a study something like mine. Moderate-sized comparative studies should be avoided. One can learn much from a single case or from comparing two cases; four cases are just enough to give the researcher some idea of the range of variation in the phenomena under study, but not enough to produce firm trends. It is not wise to postpone important decisions to the field-work stage. Extensive and detailed research decisions should be written down prior to taking to the field, and changed only after careful deliberation. In the field, the researcher will be tempted to take the advice of local sociologists, assuming that they know their local culture. This often turns out not to be the case especially when they are not specialists in the area under study. There is a mystique about comparative research which probably arises because a certain amount of sloppiness is inevitable. Comparative research is only a special form of replicative study where one must resolve the dilemma of maintaining a design for the sake of methodological consistency or changing it because of unanticipated events and thereby lose comparability of findings. I tended to opt for consistency, but am now not sure that I made the correct decisions. I did enjoy most of the enterprise, but I'm glad that it is over.

Acknowledgments

Over one thousand autoworkers invited us into their homes and did their best not to make us feel like stupid intruders. I am grateful to them above all. Cross-national research costs a lot of money. I could not have finished the research without financial support from The National Science Foundation, The Social Science Research Council, The Fulbright Fellowship Program, The Ford Foundation Fund administered by Michigan State University's International Programs, the School of Labor and Industrial Relations at Michigan State, and the Institute of Labor and Industrial Relations at the University of Illinois at Urbana-Champaign.

Professor Paolo Ammassari of the University of Rome was field director of the Turin study and a colleague on whom I placed too heavy a burden at times. Together he and I set the pattern of the future studies. I owe him more than I can repay. Professor Fillipo Barbano of the University of Turin was my friend, host, and advisor throughout my stay in Turin. The research would have come to an untimely death without his generous help.

I was fortunate to have talented assistants to direct the field work in each country. I learned more from them than they realize. Steven Deutsch did the job in Lansing; Richard Gale in Córdoba; Baldev Sharma in Bombay. Others who helped process the data and gave the benefit of their good advice include: Mohammad Salehi, Frank Holland, Noreen Dulz, Camille Legendre, Robert Bibb, and Fred Pampel.

I have profited from discussions with Alex Inkeles, William Faunce, Joseph LaPalombara, Sigmund Nosow, Bernard Karsh, Kenneth Spenner, Mauricio Salaún, Wilbert Moore, and John Pease. My most critical, demanding, and sympathetic colleague has been Joan Huber.

Finally, I am grateful to the editors of the following journals for permission to reproduce some of the materials which appeared in article form: The American Sociological Review, the American Journal of Sociology, Social Forces, Industrial Relations, The British Journal of Sociology, Studies in Comparative

Economic Development. The criticisms of the anonymous readers of these journals is greatly appreciated. Anice Birge supervised, typed, corrected, and otherwise improved several drafts of the manuscript and I am most grateful to her and her assistant, Carol Williams.

I gratefully acknowledge permission granted by the following periodicals to use materials from articles they originally published. I have cited the original articles; the chapter of the book which includes the materials is given in parentheses. With one exception, these articles have been completely rewritten to fit the overall design of the book.

"The accommodation of rural and urban workers to industrial discipline and urban living: a four nation study." Rural Sociology 36 December 1971:488-508 (Chapter 4).

"Technology and social behavior of workers in four countries: a sociotechnical perspective." American Sociological Review 37 December 1972:727-738 (Chapter 5).

"The internal stratification of the working class: system involvements of auto workers in four countries." American Sociological Review 38 December 1973:697-711 (Chapter 10).

"Auto workers and their machines: a study of work, factory, and job satisfaction in four countries." Social Forces 52 September 1973:1-15 (Chapter 6).

"Job vs. political unionism: a cross-national comparison." Industrial Relations 12 May 1973:244-238 (Chapter 8).

"Automobile workers in four countries: the relevance of system participation for working-class movements." The British Journal of Sociology 25 December 1974:442-460 (Chapter 9).

"The social construction of anomie: a four nation study of automobile workers." The American Journal of Sociology 80 March 1975:1165-1191 (Chapter 11).

"Field problems in comparative research: the politics of distrust." Studies in Comparative International Development 8 Spring 1973:20-48 (Appendix B).

William H. Form
Champaign, Illinois

Blue-Collar Stratification

TURIN CH'A BOUGIA
Canssonetta's l'aria del Sonador Ambulant di A. Dughera

Con l'invenssion d'ades
Turin l'è'n moto,
L'industria fa progres,
Tuti lo noto,
Sarà 'n cost'occasion
Un grand afè
Per chi professa n'arte
O un mestè.

Trista la vita, sempre gumè,
Travaié sempre e mai gnun piê.
Ma son l'è niente, s'a iè d'afè
Piand quaich sborgne tut fa passè.

Pensand i'ani 'ndare
Dio che crisi!
Si sa che nen da fè
L'era 'n suplissi.
L'industria a nost pais
L'era 'ndarè.
Bohèm as na fasia
'N tuti i mestè.

Trista la vita, sempre gumè, ecc.

Giá, l'epoca d'ades
Son maravie,
As fa dij gran succes,
lè nen da rie.
con tute côste grandi
Invenssion
A n'aria, as dominrà
Fiña 'l balon!

Trista la vita, sempre gumè, ecc.

A gara a van ie sgnôri
e i grandi nobii
A andé pi che 'l vapôr
Su l'automobij,
A van che, già as capis
A fan di guai.
L'è certo per la FIAT
Un gran traval.

Trista la vita, sempre gumè, ecc.

TURIN ON THE MOVE
Song's Melody
by Troubadour A. Dughera
Turin, 1907
Translated from
the Piemontesi
by William H. Form

With the inventions of today
Turin is on the move,
Industry is booming,
Everybody knows it,
In this situation
There'll be big things coming
For whoever practices a skill
Or a craft.

Life is sad, always troublesome,
Always working but earning nothing.
But this is nothing, if they're troubles
With a few binges, everything passes.

Thinking about past years,
God, what crises!
Certainly, without work
It was pure hell.
Industry in our country
Was really retarded.
Nobody was well off
Whatever his occupation.

Life is sad, always troublesome, etc.

But nowadays
Things are astonishing,
With such great achievements
There's little to ridicule.
With all these huge
Inventions in the offing,
They'll even surpass the balloon!

Life is sad, always troublesome, etc.

Lords and nobles compete
To travel more by auto than by ship,
Carrying on this way, they'll sure
Make trouble.
But certainly for FIAT
They'll bring a lot of business.

Life is sad, always troublesome, etc.

1

Technology and the Social Integration of the Working Class

When England, Germany, the United States, Japan, and the Soviet Union industrialized, they created large classes of factory workers who proved to be difficult to manage. No matter which elite directed industrialization, factory workers were poorly paid (economically exploited) because surplus capital was needed to build more, larger, and more sophisticated factories. Factory managers tried to legitimate their policies in the name of human nature (greed), religious virtue, state glory, or, ironically, a utopian future for the workers themselves. These attempts were not permanently effective because workers resorted to sabotage, slowdown, strikes, and even revolution to get a larger return from their efforts. In these situations, management's political formula was to make just enough concessions to assure loyalty to the system. But the system has usually been turbulent. Consequently, managers, political leaders, and social scientists have been haunted by the question of whether factory workers would strike out for themselves or whether they would continue to go along with the system.

If the social sciences have agreed on anything, it is that the social cement of traditional societies has been eroded for good or ill by changes emanating from the factory system. Optimists like Marx and Durkheim believed that technological change would eventually produce a new society which would integrate workers into it, but Weber thought that bureaucratic and rational revolutions would reduce societal solidarity. Though divided on the future, theorists have agreed that the shift to industrial society was especially painful for the working class. Hundreds of scholars have commented on the disorganized, segmented, anomic, alienated lives of industrial workers. Like their medieval predecessors, modern scholastics argue endlessly about what these terms "really" mean.

A strong subjective humanist orientation pervades much of the literature on industrial change because social scientists are ambivalent about machines and the factory system. While they applaud the material benefits of the machine age, they deplore the fact that somebody has to tend the machines. Most social scientists, with a simplistic view of the factory as represented by the automobile assembly line, chant a seemingly endless litany on how machines alienate workers from their work, work-groups, societies, and even themselves. The chanting seems inspired by hopes that workers will galvanize their unions to launch political movements which will alter not only factory life but the entire political economy. The spread of industrialism to "underdeveloped" countries is seen as creating an even vaster pool of alienated workers who will join Western workers to bring about a more just world civilization.

While few can argue the need to improve factory life and the desirability for greater equality in income and power in society, little is gained by substituting ideology for empirical studies. Most social scientists have a mushy view of modern technology and how workers respond to it. Either they see technology in Chaplinesque terms as a man (never a woman) so tied to a machine that he cannot make an independent move or they see it as a global ooze enveloping everyone and everything in its path. But modern technology is enormously complex, and human responses to it are just as complex. Even in mass-production industry, industrial employees are not mass men or women. The concept of a mass citizen denies workers their dignity, stereotypes their behavior, and fails to take into account their changing responses to change. In short, I am skeptical of dominant social science views of how technology affects workers' lives, and I shall reexamine these views in a cross-national context.

Three assumptions underlie the research. First, workers are not socially homogeneous before they enter the factory; second, factory technology has variable effects on the social life of the worker in the plant; third, different backgrounds and technological relations of workers produce different participation profiles in the community. In the industrial sector of any society, the most important socialization variables are the educational and occupational backgrounds of the workers. In the factory, the most important factor differentiating social life is the occupational skill of the worker. In the community, the most important differentiating factor is organizational participation. The task of this research is to examine how social background, skill, and community ties relate to each other in the social life of workers in countries which differ in industrialization. Several important questions will be studied.

How great are the stratal differences among workers prior to factory employment and do they change with increasing industrialization of the country?

Does the technology of the factory (in terms of its effects on workgroup formation) undermine or strengthen stratal differences among workers in different countries?

How do different technological relations affect the worker's involvement in the union and do different political ideologies affect this involvement?

How do different work-group and union relationships affect worker participation in the neighborhood, community, and nation?

Does increasing industrialization solidify or stratify the organizational lives of workers inside and outside the factory? What are the implications for working-class social movements?

These important and complex questions cannot be answered by a single study. I have therefore tried to simplify the attack on these questions by studying a single technology which is highly standardized (automobile) but found in countries varying in industrialization. Variation in the technology is assured by selecting in all four plants occupations with different skill demands: assemblers, operators of semi-automatic machines, inspectors and related occupations, and skilled-craft workers. To consider whether technology has an effect independent of its broader industrial milieu, I chose countries with varying levels of industrialization: India, Argentina, Italy, and the United States. A wide range of social systems are examined: family, work group, union, friendship networks, neighborhood, community, and the nation. Finally, I reduced problems of interpretation by accepting what workers say about themselves rather than the views of their self-appointed spokesmen.

Three Approaches to the Study of Industrialization

Technology is the application of nonhuman energy for human purposes. Machines constitute the main nonhuman energy converters in industrial society, but to think only of machines as technology is too simple. In an advanced industry such as automobile manufacturing, many different machines are linked into a vast occupational/organizational system which constitutes the plants' technology; each machine requires particular occupational skills, and the linking of the machines requires a certain type of factory organization. The ecology of the factory (machine placement), its management organization, occupational structure, informal organization, union organization, and many other features make sense only in terms of a particular technological system, but this system does not bear down on all workers alike. My task is to unravel the complicated ways the technological system affects worker behavior and the various ways workers respond to it inside and outside the factory. Although this is an ancient social science question, the industrialization of nonwestern countries today calls for a reexamination of old explanations.

Historians once agreed that, before the advent of the machine revolution, societies were highly integrated, but population surpluses in the countryside and high demand for factory workers stimulated long-term migration to the city. As the economic basis of traditional societies was eroded by new market dependencies, the stratification systems of both rural and urban communities changed. In the city, an occupationally based class system arose, at the bottom of which were factory employees, the so-called working class. This class suffered most from social change because it was cut adrift from its rural heritage and urban institutions had not yet evolved to handle such problems as unemployment, disease, bad housing, and family breakdown. In response to many pressures, urban organizations (welfare, health, education) arose to meet some of the most pressing problems. Factory workers were slow to organize permanent mutual aid organizations to look after their interests. But as they gained experience in how to handle conflict with management, they formed labor unions. Later recognizing that only limited gains could be achieved by collective bargaining, unions tried to expand working class influence by forming separate labor parties or political bodies. Although labor became a part of the political community, it rarely made major economic decisions, and technological change continued to affect it. Today, many industrial workers feel that they cannot shape their work nor influence the main institutions of society (Bauman, 1972). But the aristocracy of labor feels satisfied with its work and societal arrangements.

After World War II, when many new nations launched programs of industrial development, social scientists saw an opportunity to test ideas concerning the way an emerging working class responds to industrial growth. Some expected European experiences to be repeated elsewhere, but this did not typically happen. Today, no completely satisfactory theory of industrialization exists, but three theoretical frames of reference guide most of the research. The "cultural" heavily influenced by anthropology, emphasizes that invading industry must accommodate to traditional cultural patterns; the "industrialism" hypothesis predicts a quick societal accommodation to industry everywhere; and the "development" or evolutionary frame of reference emphasizes that a limited number of institutional responses are possible at different levels of industrialization. The industrialism and development perspectives both stress the independent power of industry to force societal changes, but the development perspective places more emphasis on the different adaptations which societies make at various stages of industrialization. This study generally follows the development perspective, borrows heavily from the industrialism hypothesis, and occasionally relies on the cultural approach to explain specific events.

Before describing and evaluating each theoretical perspective, we may reasonably ask what differences it makes whether one follows one rather than another. Using the issue of the development of a working-class social movement as an illustration, the cultural perspective is conservative; it stresses that social movements will fit traditional power relationships. In Japan, for example, enterprise unionism is a response consonant with the paternalistic patterns in industry which flow from a persistence of feudal relations between owners and employees. An independent working-class movement is unlikely to appear. The industrialism perspective is more radical. It emphasizes the inevitability of conflict between managers and employees whatever the earlier stratification system and the present political economy. Unless suppressed by a totalitarian government, a working-class movement appears quickly. Japanese, American, and Indian unions, according to this view, are more alike than different because they are basically concerned about the distribution of power in the factory and in the society. The development perspective stresses that certain conditions in society must develop before a working-class movement is possible. When present, aggressive politically oriented unions can lead the movement. But if they do that, a separate working-class movement may not emerge as workers become immersed in a system of pluralistic politics. Each theoretical perspective will be described and evaluated.

Culturologists hold that indigenous cultures can block industrial invasion, but if they do not, they force industry to accommodate to traditional institutions. Few social scientists follow this view rigidly, but those who do recognize that cultures may vary in their ability to resist industrialism, but everywhere industry must make some accommodations to local culture. Workers do not become disciplined and committed employees overnight, and many political, religious, and other elites resist Western practices and bend industrial systems to harmonize with their interests and values (Nash, 1958; Kriesberg, 1963). Japan is often cited as a society which forced the factory system to accommodate to its feudal institutions (Abegglen, 1959). Bennett and Ishino (1963: 343) point out that, though paternalism is often practiced in factories of societies at early stages of industrialization, the pattern differs among Latin American, Japanese, and other societies. In Japan, a fictive hierarchical kin ideology defines the relations between supervisors and workers, while in Latin America the patron system is carried over in the factory. Okochi *et al.* (1974) point out that Japan is the *only* advanced industrial society whose stratification system is based on traditional status relationships and not upon an occupational hierarchy. The cultural perspective is often espoused by social scientists who are area specialists or experts in the social organization of particular tribes, peoples, or regions (cf. Nakane, 1970).

In its extreme form, the cultural perspective is of little utility because it denies the possibility of making transcultural generalizations, a position I cannot accept. The waning of tribal cultures, the passing of peasant societies all over the world, and the spread of the factory system makes the cultural perspective hard to defend. While documentation of how individual factories adapt to local cultures in the early phase of industrialization is valuable ethnographic data, many scientists want to generalize before all the data are in. Yet the cultural approach is needed to meet some problems. In this study some interview questions were modified in each nation to take into account unique cultural meanings. The neighborhood idea presumably does not exist in India. Yet I was interested in local urban relationships, so a conceptual equivalent for the neighborhood had to be found (cf. Straus, 1969). In Argentina, the involvement of workers in the union had to be interpreted in the context of Peronism in the labor movement. In short, culturally relevant data were needed in all countries to interpret some findings.

The industrialism approach is generally known as the Industrial Man hypothesis, a designation I reject because of its sexist bias. Incidentally, in all four countries no women were employed in the departments I studied, an interesting cross-national instance of sex stratification. Kerr and his associates (1960) first elaborated an institutional expression of the industrialism hypothesis, and Inkeles (1960) elaborated its social psychological expression. In both views, the introduction of the factory everywhere shatters traditional institutions and changes workers' perspectives of the world. In the Marxist tradition, both hold that the factory system produces uniform material products, similar institutions, and common human responses. Kerr and his associates note that the organization of the factory is similar everywhere, especially for the same industry. For example, variation in the organization of a steel mill with a given technology is very limited. Moreover, all societies must build institutions that accommodate to the factory: workers must be trained for new jobs, employment bureaus must gather job-market information, organizations must handle mass unemployment and retirement problems, and agencies must settle disputes between labor and management. While the organizational forms devised to provide these services may vary from one society to another, they are limited in number. Labor disputes which inevitably arise must be handled either by a factory department, a labor-management body, or a government bureau. Finally, the factory system exerts powerful secondary effects on social organizations not directly tied to it, such as family, market, stratification system, neighborhood, community, and government. This dynamic industrial pattern (factory system, associated organizations, and societal accommodations) will spread and eventually produce a single world society.

Inkeles demonstrated that these organizational changes have their social psychological counterparts. Regardless of culture, factory technology quickly disciplines employees to display a set of common attitudes, values, and beliefs. The factory, like the school, is an educative force, producing people with a modern mentality; they are more rational, believe in formal education, abandon traditional loyalties, and develop wider world views. In short, both the institutional and social psychological views emphasize the power of industrialism to overcome the resistance of traditional cultures.

In an effort to modify extreme positions, Gusfield (1967) stressed that many elements of traditional societies are compatible with industrial demands, that traditionalism and modernity are misplaced polarities, and that conflict is not an inevitable response to the factory system. He believes, for example, that Hindu philosophical and religious teachings are compatible with several life orientations, including the industrial.

The industrialism hypothesis is useful for this study. It predicts, that since all four factories have similar technologies, they should have similar management organizations, occupational structures, labor problems, pay scales, and many other features. My research did not call for studying the second order organizational consequences of industrialization: e.g., how the family, stratification, educational, and other systems respond to the factory. But Inkeles' hypothesis that industrial workers everywhere should respond similarly to common structures could be tested. I expected employees (undifferentiated by skill level) in the four countries to respond similarly to their jobs, working conditions, union programs, and other matters. However, the hypothesis is too loosely stated to offer guidance on how much similarity should be expected in attitudes, beliefs, and values, and it does not attack the problem of how workers will become involved in work groups, unions, political parties, community organizations, and working-class movements. Nor can the hypothesis explain differences in worker responses when exposure to industry is held constant.

The third position is known as the developmental, evolutionary, or convergence hypothesis (depending on the emphasis), and postulates that societies at various stages of industrialization face different problems, but as they become more industrial, their structures become more alike. Although Moore (1965) and Smelser (1963b), its major theorists, are receptive to some features of the industrialism hypothesis, they believe that industrial development is a complicated process that moves irregularly in response to various social structural conditions. For example, India and Argentina are both industrializing, but they already differ in industrial makeup, the problems they face, and their trajectories toward industrialization. While upper strata in both societies have resisted the factory system, they have responded differently to it; many Brahmins have become factory workers but few upper-class Argentinians have done so.

Karsh and Cole (1968), Taira (1970), and others contrast the American and Japanese cases to demonstrate different paths in the structural convergence of industrial societies. While they acknowledge some unique adaptations of Japanese institutions to industrialism, they feel that current changes lead toward the industrial pattern found in the West. Lifetime employment has been practiced by many large Japanese enterprises, along with a wage scheme based on status variables such as age and length of service. However, newer industries are resisting the institutionalization of lifetime employment, and young educated workers are insisting that they be paid according to their skill, as in the United States, and not according to their status.

The development frame of reference is not without limitations. Perhaps its least tenable assumption is that scholars know the path of development: e.g., that the United States represents the most highly developed industrial nation. Dore (1973) asserts that Japan has jumped stages of development and that her institutions are better suited to an industrial society than are Western institutions. It is the West which must develop toward the Japanese model. Similarly, Armer and Schnaiberg (1973:277) suggest that industrial development and modernization are ethnocentric concepts with political overtones, and perhaps should be abandoned.

Testing the development hypothesis requires historical data from many societies which vary in their levels of industrialization and in their cultures. No one has been able to test the hypothesis fully because we do not have a sufficient number of societies with different cultural heritages at all levels of industrialization. Typically, the hypothesis has been examined piecemeal. Thus, Bendix (1956) examined changing managerial ideologies during the industrialization of Western societies; Harbison and Burgess early (1954) studied management behavior in European societies; Cole (1971) examined worker relations in a Japanese factory; Lambert (1963) studied the behavior of Indian industrial workers. Many of these studies are not systematically comparative (cf. Marsh, 1967); a problem is systematically studied in one nation and the data are compared to whatever data are available on other nations.

For the sake of parsimony, I have exaggerated the position taken by scholars following the three theoretical perspectives. Most of them combine the three approaches, and emphasize one of them. This research follows the development hypothesis, but leans heavily on the industrialism hypothesis and occasionally calls for a cultural explanation. In this study technology is treated both as a constant[1] and as a variable. Automobile technology represents the constant, and extent of national industrialization represents a contextual variable of points along a continuum of industrialization (cf. Przeworski and Teune, 1970:17-30).[2] India represents early industrialization; Argentina and Italy, moderate development; and the United States, an advanced industrial society. The theoretical problem is to generate hypotheses about system involvements of workers which take into account extent of industry in the country as a contextual variable. Other variables must also be taken into account. Thus, the labor unions of the four countries differ in organization, ideology, and political ties; the ecology and institutional arrangements of the cities selected for study differ; and the nations differ in stratification patterns, political institutions, and the problems they face. Some substantive knowledge is required of these systems in order to interpret the findings.

This study has modest ambitions, and does not test the development hypothesis directly; rather, it searches for commonalities and differences in workers' system involvements which can be explained by a few principles of technology and stratification. The thrust is not to explain how these systems function, but how workers become involved in them. The development hypothesis is not a fully developed theory with a set of propositions which explain how advancing industrialization affects social system involvements of workers. Three important issues dominate the development literature. The first is whether advancing industrialization affects workers' responses to their jobs and the factory social system. Do they hate their work more with advancing industrialization and become more dissociated from their fellow

[1] Technological differences among the factories and skill levels are treated as variables for some problems in Chapter 5.

[2] Comparative research has high currency among social scientists, but I prefer to think of this study in the replicative tradition: the same problem is studied with the same method in four countries which differ in industrialization and institutional pattern (cf. Finifter, 1973).

workers, or do they become more adapted to industrial society? The second issue concerns the response to unions. Does industrialization encourage workers to become more militant or do they settle for "mature" labor-management relations? The third issue concerns the organizational environment external to the factory and union. Do workers become increasingly self-conscious as a class and link their organizations to fight for their interests or do they become absorbed in mainstream organizations and politics? In some societies this is the issue of proletarianization versus embourgeoisement (Goldthorpe *et al.*, 1969). Some students hold that generalizations cannot be made on these three issues because workers in different occupational strata respond differently to them.

These three issues are difficult ones and the data from the four cases of this study cannot settle them. In its extreme form, the industrialism hypothesis minimizes differences among industrial nations and points to the United States as a case where the three issues are settled: workers quickly adjust to their jobs and build satisfactory relationships with their fellow workers; union-management relations mature quickly and workers push their unions to institutionalize the settling of differences; workers are apathetic about the class issue and they are pulled into the middle mass of society.

The American Case

Many scholars object to these conclusions as a conservative view of the American scene which is inapplicable to other countries. I did not expect the American data to be replicated in detail in the other nations, but I did expect some principles which explain the American findings to apply elsewhere. I will first present a brief summary of the vast American literature and then focus on the principles which explain the findings. These principles are basic to my views on the relationship between industrial development and worker system involvements.

Despite the negative image that social scientists have of industrial work, most American studies show that employees do not try to escape the factory and that they find satisfaction even in the most routine jobs. New industrial workers accommodate quickly to industrial discipline, managerial authority, and union organization. They report satisfactory relations with their workmates both inside and outside the factory and, although they rarely attend union meetings, they support the union's attempts to raise wages and improve working conditions. But they do not like the union to use their dues for political campaigning and other purposes. The personal bonds which workers build in the factory and union do not carry over into the community; their social life centers on family and neighborhood activities. Though few workers are members of local and national organizations, they are moderately well-informed about the major community and national problems which affect their economic interests. Yet they feel that workers and their unions have little political power. Finally, workers are aware of the stratification system and their place in it. Though dissatisfied with their present status, they are not anxious to change it through political action, and they hope that their children will escape factory jobs and become professionals.

The dominant image of the autoworker is the man on the assembly line, though less than one-quarter have that job. Stratified by skill level and function, autoworkers are not a homogeneous mass. Compared to most factory workers, the skilled are recruited from privileged backgrounds and their advantages improve over time. Craft workers have more freedom on their jobs and more control over their work, and they are happier at it. They form more solidary work groups, participate more in the union, and tend to control it. The skilled prefer the union to bargain with management rather than engage in militant confrontations. Outside the factory, the skilled participate in more community and national organizations than do other workers. Craft workers are more politically active and more conservative than the less skilled. Aware of their interests and advantages over the less skilled, craft workers fight to maintain their position, thus threatening the solidarity of the working class. The literature is inconclusive on their class sympathies, but recent research suggests that they see themselves as an independent stratum (Mackenzie, 1973).

Some Explanatory Principles

How can these findings be explained? The market situation and status of first-generation unskilled industrial workers are different from those of skilled workers, who often are second-generation urbanites. Whatever their previous occupation (usually they are rural migrants or low-status workers in the service sector), new factory workers find industrial work attractive. If they are given the choice between becoming factory workers who receive regular wages which can be used to buy goods and services, or remaining in self-sufficient agriculture (or self-employment in a marginal business) factory work wins hands down. For most first-generation factory employees, a factory job, especially in the high-paying automobile industry, represents upward mobility.

Exhausting and dull work is a way of life for most rural workers, and dirty, demeaning work is common in the urban service sector, especially for women. In contrast, factory work may be less physically demanding, less lonely, less dirty, less demeaning, and even less routine. Moreover, work and job satisfaction have low salience for new industrial workers. They prefer a steady, even low-paying factory job, however dull or dirty, to their previous jobs in the agricultural or service sector. They prize economic security so much that they suppress traditional religious, ethnic, and other hostilities which may arise in the factory.

New industrial workers typically accept industrial organization, managerial authority, and union jurisdiction because they have no choice; nor have they a separate solidary organization (class, tribe, kinship network) to challenge factory organization. Unions are accepted as part of the factory, almost like management. But since the political activities of union officers appear unrelated to work grievances or income goals, workers are unenthusiastic about political unionism. Even solidary work-groups rarely influence neighborhood and community activities, because their members are dispersed residentially. Low organizational participation is to be expected on several grounds: workers are new to the city, participation costs money, and organizations are already controlled by higher-status groups. Moreover, new workers see the occupational structure of the factory as reflecting the stratification system of the community and they recognize their place in both stratification systems. The prevailing ideology informs them that they are not trained for higher positions, but that their children may become educated for them. Since workers are aware of their political powerlessness and the inconsistency between their low status and the mobility ideology, they view the society as more disorderly (anomic) than do people in higher occupations.

This profile of worker behavior and social involvements describes a wide range of cases, with three exceptions: second-generation industrial workers, factory employees in small communities, and workers in societies with planned economies. While some differences surely exist among new industrial workers in societies of high and low industrialization (to be described later), their similarities are strong the world over.

The different system involvements characteristic of skilled and unskilled workers may be explained by a few principles of technology and stratification. Since skilled workers are recruited from more privileged backgrounds, they receive more formal education and technical training, which in turn leads them to higher-paying jobs. Maintaining these advantages becomes obsessive, so they seek influence in the work-group, union, and associated organizations. Their success may be explained by the interaction advantages they have on the job. Since they have direct control over their work, they can move about at will and interact with their workmates. Consequently, the skilled have more solidary work-groups than do the less skilled, an advantage which they use in informal bargaining with management and in struggles to control the union.

Crafts workers often are second-generation urban dwellers who know how organizations function. They attend union meetings more frequently than the less skilled, become union officers, and use union power to extend their advantages. The union's ties to other unions and other organizations, such as political parties, extend the organizational horizon of the skilled; they participate more in politics and organizational activities which affect their welfare. The more solidary the work-group, the more it influences its organizational environment. For the skilled, organizational participation is both a source of influence and a consumption activity. In sum, advantages derived from job technology and position in the stratification system give the skilled many advantages over the less skilled. Superior background, better work, more pay, and more solidary work-ties increase their bargaining strength in the factory, union, and community. Finally, the superior ability of the skilled to link their social systems and their broader organizational horizon induces them to see society as relatively organized and integrated.

While the technology of the four factories under study is similar, it is more complex in the most industrial societies. The more machines differ in complexity, the more differentiated are the opportunities for work interaction. Differences in interaction opportunities among the skills are therefore greatest in the most industrial societies. Skills are less differentiated in underdeveloped societies because workers operate simpler machines. It follows that interaction opportunities of the various skill levels tend to be more alike. Moreover, factory managers in less industrial countries blur status distinctions among the skills by paying workers on the basis of status (e.g., length of employment and loyalty) rather than skill and performance. In sum, the complex technology of highly industrial societies and their adherence to universalistic norms of job payment serve to differentiate interaction rates among the skill levels and formalize their income differences, thus increasing the stratification of the labor force.

Industrialization and System Involvements[3]

The straightforward application of the industrialism hypothesis emphasizes similarities in system involvements among workers in the four countries. But the development orientation calls for demonstrating how system involvements (kinship, friendship networks, union, neighborhood, community, and nation) change with increasing industrialization. Since literature on this subject is scanty, I shall suggest some possible relationships between industrialization and system participation and reserve qualification for later chapters.

The technological level of societies is often measured in terms of the amount of energy people use per capita: United States is highest, India is lowest, and Italy and Argentina fall in between.[4] Technological level and organizational complexity of societies typically are highly related. Organizational complexity has two

[3] I do not like to use the concept of social system because of its close association with functional theory, but I cannot think of another term which embraces the wide range of entities in which I am interested: kinship, friendship networks, union, neighborhood, community, and nation.

[4] Differences among the four cities are smaller than differences among the countries, but the order of industrial complexity remains.

major dimensions: organizational density (number of organizations per thousand residents) and network density (number of interorganizational ties per thousand organizations). Increasing organizational and network densities increase the opportunities for organizational involvement and decrease ties to ascriptive groups such as family and kin.

Industrialization changes family interaction by changing its organizational environment. In early industrial societies, the family absorbs most of the nonwork time of employees because the external environment is simple; specialized economic, educational, recreational, and other organizations have not yet developed on a mass basis because they are costly. But advanced industrial societies create a complex organizational environment which involves individual family members differently: parents may work in different factories and belong to different unions, children may attend different schools, and individual family members may participate in different recreational organizations. Industrialization thus reduces the family and kin contacts of workers and differentiates their activities with friends, workmates, and members of common organizations. The salience of nonfamily contacts for the formation of economic and political beliefs increases as family interaction declines. The higher the workers' skills, the more salient are nonfamily contacts for political socialization. While this holds irrespective of societal industrialization, it is truest in the most industrial society. Paradoxically, union and political affairs are less important family topics where industry is most highly developed, decreasing the role of the family in political socialization.

Neighborhood involvement influences working-class solidarity if the bonds forged in the work-place are reinforced by neighborhood contacts. The more neighborhood contacts are organizational, the more important they are for working-class movements because organizations can be linked to networks in the external community. The more that economic services (markets, bars, movies) and institutional services (educational, religious, governmental) are available locally, the greater is the potential for neighborhood involvement. Since the decentralization of specialized market and institutional services into neighborhoods is costly, they tend to be concentrated in the city center in countries where industry is undeveloped. However, since effective demand for these services is low and transportation to the city center is costly, workers' contacts in the neighborhood remain almost completely interpersonal and have little relevance for political involvements.

In the moderately and highly industrialized societies of Latin America and Europe, markets and institutional services are generally available in urban neighborhoods, reducing the necessity to leave them except for highly specialized needs. Relatively inexpensive public transportation is also available. In addition, neighborhoods tend to be large, have names, and serve as legal entities of the city. The overall result is high neighborhood involvement. When neighborhoods are occupationally homogeneous and occupied by second generation industrial workers, working class solidarity tends to be even higher.

The United States represents a special case. The growth of cities during the automobile era dispersed special economic and institutional services over wide areas. But the widespread availability of automobiles made these services available regardless of their location. Today, almost all economic services, including food stores, are located outside the immediate neighborhood, as are institutional services, with the exception of grammar schools. These conditions, plus high residential mobility, decrease neighborhood involvement and local consciousness. In sum, the neighborhood involvement and network density necessary for developing working-class consciousness should be high in moderately industrialized countries, lower where industry is undeveloped, and lowest in highly industrial countries which have highly mobile residents. Although residential segregation along skill lines is not marked among manual workers, skilled craftsmen should be less involved in neighborhoods than the unskilled because the skilled can afford to pay for more services which pull them out of the neighborhood.

The labor union is pivotal for the worker's organizational involvements because it addresses both plant and outplant concerns. Those of the plant deal with working conditions and labor-management relations, while outplant concerns focus on political parties and governmental operations. Increasing worker-union-party-government linkages do not necessarily accompany industrial development, but organizational linkages of the union to government tend to be stronger in the less industrial countries, and worker participation in union activities tends to be more widespread in highly industrial countries. Industrial growth is of indirect causal importance. With important exceptions, economic policy formation by government tends to be more highly centralized in less industrial countries; union leaders think that they can improve the well-being of workers more quickly by influencing governmental policies than by focusing on plant issues. But continuous technological changes in the factories of the more industrial countries force workers and union officials to pay more attention to changing working conditions and productivity rates. At the same time, union officials urge their members to become involved in political and governmental concerns. The overall result is that workers become highly involved with working conditions and productivity issues but only moderately involved with political action. In contrast, in the less industrial countries, the union tends to neglect plant issues and convinces only a few employees to become involved in political activities.

The union's links to nonpolitical community and national organizations (educational, consumer, medical, welfare) is more highly developed in the denser organizational environment of the highly industrial countries. These contacts are almost monopolized by a few skilled workers and union officials. In the less industrial countries community involvements of the union and its members are negligible. In short, all workers, whatever their union ideology, are de facto bread-and-butter unionists, but in the more industrial countries they are involved in a broader range of union activities inside and outside the plant. Here the possibility of building working-class solidarity is high if workers are not split by different occupational and community interests.

Worker involvement in the community and nation are considered together because most participation occurs locally. The more industrial a society is, the more workers are involved in larger systems and the more they know about them. But industrialization is not the only cause; rising educational attainment stimulates the consumption of information about the community and nation.

The organizational society confers membership on its citizens almost with ascriptive insistence; membership in union, PTA, bowling league, neighborhood association, political party, or religious society is almost a by-product of having a job, having children, living in a neighborhood, or having a religious preference. The pattern of increasing consumption includes consumption of information; channels of communication (newspaper, magazines, radio, movies, television) multiply, and workers are more exposed to information about the wider world. Even ordinary consumption, such as buying an automobile or acquiring a new hobby, extends workers' interpersonal, organizational, and communication networks. In sum, even though workers everywhere belong to relatively few organizations, their exposure to organizational information increases with industrialization. The more skilled they are, the more they are pulled into the broader community and the less they are involved in the neighborhood and family. Unless their community and national contacts are dominated by the union, workers tend to become cosmopolitanized and less absorbed into working-class culture. The more skilled they are, the more this is the case. Moreover, the amount of societal anomie workers perceive probably decreases with industrialization, both because the society's organizational problems are better resolved and because workers are increasingly absorbed into its organizational life.

The development orientation suggests that the social-system involvements of industrial workers are due not to national or cultural differences, but to the technological and organizational complexity of the industrial sector and to differences in their skills. Current notions that workers are increasingly dissatisfied with their jobs, that political unionism grows with industrial development, and that workers feel increasingly alienated from the factory, union, community, and nation represent an oversimplification of reality. The theory offered here suggests that the working class becomes less homogeneous with industrialization, that it becomes increasingly stratified internally, and that there is no inevitability in the political patterns which emerge.

I have not considered how the technology of the post-industrial era will affect the work, union, and community involvement of workers (Bell, 1973). However, even in the United States, the overwhelming majority of the labor force is employed in organizations which use only mechanical technology (Parker, 1972). Automation is not proceeding as rapidly as early observers predicted, and only a tiny segment of the automobile industry has been automated. Moreover, recent research has shown that the impact of automation on the skill-composition of the labor force is small. When a new technology emerges, engineers tend to classify workers who deal with it as skilled. After training for the technology is routinized, skills are downgraded, a phenomenon presently occurring in both the factory and the office. Fox (1974: 344) examined studies on automation effects on job skills and concluded that the effects are mixed and that early predictions were exaggerated. I am agnostic about the ultimate impacts of automation, but I do not believe that the failure of this study to consider them affects its general conclusions.

This book is divided into three parts. The first describes the industrial development of the nations and cities selected for the study, the second examines the effect of technology on worker behavior in the factory and the union, and the third discusses the effects of industrialization on worker involvement in nonplant social systems. Background data in Part I are needed to answer two major questions: first, whether the working class is stratified prior to entry into the factory, even in countries of low industrialization, and second, whether exposure to rural socialization, which varies enormously in the four countries, affects worker accommodation to factory and urban life. Part II pursues the stratification theme by considering whether the technological environment of the factory affects the work-group solidarity of the various skills and their struggle to control the union. A parallel question is whether industrialization increases job dissatisfaction and the pressure for militant and ideological unionism. Part III asks whether stratal differences in the factory find expression in different patterns of community and national involvement. The final question is whether industrialization stimulates the internal stratification of the working class and the development of successful working-class movements.

2

Four Nations, Four Cities, and Four Factories

India, Argentina, Italy, and the United States differ considerably in extent of industrialization; Bombay, Córdoba, Turin, and Lansing differ less, and the factories (PAL, IKA, FIAT, and OLDS) are somewhat alike. A basic research question is whether the different national and urban environments in which industrial workers live, affect their life patterns. The first task is to demonstrate that the national and urban environments were, in fact, different. Table 2.1 which presents data on social and economic

TABLE 2.1 INDICES OF SOCIAL AND ECONOMIC DEVELOPMENT

Factors	*U.S.A.*	*Italy*	*Argentina*	*India*
Percent in manufacturing (1960)	33	40	34	12
Per capita consumption[a]	8,013	1,186	1,069	140
Per capita GNP, U.S. dollars[b]	2,577	516	490	73
National development rank[c]	1	23	27	96
Human resources development index[d]	35	57	82	261
Percent in cities over 20,000[e]	52	48	30	12
Percent population literate (1960)[f]	98	92	92	28

[a] United Nations Statistical Yearbook (1961:278-279).
[b] Harbison and Myers (1964:42-48).
[c] Farace (1966:305-313).
[d] Harbison and Myers (1964:47-48).
[e] Russett *et al.* (1964).
[f] U.S. Bureau of the Census (1971:162).

indicators for the four nations, clearly demonstrates that the United States, Italy, and Argentina are relatively close together on all indicators compared to India, which is a clear case of low development. Yet Italy and Argentina rank sufficiently below the United States in per capita GNP and consumption to represent middle range cases of industrial development. Although Italy ranks higher than Argentina on the important indicators of per capita GNP, consumption, and especially in human resource development (technical occupations and advanced educational training), the differences between the two nations are relatively small.

Perhaps the level of industrial development of the city and the state, rather than that of the nation, affect the way workers live. After all, workers live in the city, which may be highly industrialized even though the country may be technologically backward. Bombay is such a case. Unfortunately, social and economic indicators of industrial development have not been worked out for individual cities and provinces, but the sector composition of the local labor force may provide a clue. Thus, Michigan employment was overwhelmingly nonagricultural, that of the provinces of Piemonte in Italy and Córdoba in Argentina was predominantly nonagricultural, while Maharashtra, India, was heavily agricultural. Yet Maharashtra was the most urban and industrial state of India, producing over one-quarter of the total manufacturing output of the nation. Piemonte was more industrial than Córdoba because five-tenths of its workforce was employed in manufacturing compared to three-tenths for Córdoba (Table 2.2). In the nonagricultural sectors, the higher the proportion of workers employed in service, the more advanced the economy is considered to be. Data in Table 2.2 show the highest proportion in Michigan and the lowest in Maharashtra, while Córdoba was higher than Piemonte. But it would be incorrect to conclude that Córdoba was more highly developed. In the more advanced technologies, a higher proportion of service workers are engaged in professional and administrative service, transportation and communication, and clerical and sales, and a lower proportion is employed in personal and domestic services. Three-tenths of Córdoba's labor force was engaged in personal or domestic service or had casual jobs in the service sector, in contrast to one-tenth for Piemonte. Piemonte had a significantly higher percent of its service workers in commerce, transportation, communication, and administrative service. In short, Piemonte ranked higher than Córdoba in the technological sophistication of its manufacturing and service sectors.

Unfortunately, adequate data on the occupational composition of the four cities were not available. In sector composition, Turin had the largest proportion employed in manufacturing, followed by Bombay, Córdoba, and Lansing (see Table 2.2). Two indicators of technological complexity were available: average number of employees per manufacturing plant and educational level of the labor force. Lansing's manufacturing plants were the largest, with a mean of 108 employees, followed by Turin (79), Córdoba (11), and Bombay (20). The figure for Bombay should be much lower because the Census reported data only for factories which hired ten or more employees. Educational level of the labor force followed the same order: the mean for Lansing was twelve years; Turin and Córdoba, between five and six years, and three years

TABLE 2.2 SECTOR COMPOSITIONS OF NATIONS, STATES, AND RESEARCH SITES (PERCENTS)

	Countries			
Sectors	*United States*[a]	*Italy*[b]	*Argentina*[c]	*India*[d]
Agriculture	8	29	21	72
Manufacturing	33	40	34	12
Services	59	31	45	16
Total	100	100	100	100
	State or Province			
	Michigan	*Piemonte*	*Córdoba*[e]	*Maharashtra*
Agriculture	4	22	25	72
Manufacturing	44	51	31	12
Services	52	27	44	16
Total	100	100	100	100
	Cities			
	Lansing	*Turin*	*Córdoba*[f]	*Greater Bombay*[g]
Agriculture	1	1	1	2
Manufacturing	31	61	35	43
Services	68	38	64	54
Total	100	100	100	100

[a] U.S. Bureau of the Census (1960).
[b] ISTAT (1964).
[c] Consejo Nacional (1965).
[d] Kulkarni (1968).
[e] Dirección Nacional de Estadistica (1960).
[f] Dirección General de Estadistica (1967).
[g] Government of India (1962).

for Bombay. In sum, for all three units (nation, state and city), the same order of technological development appeared: the United States was the most developed, followed by Italy, Argentina, and India. Italy and Argentina were closer together than any other two nations, but Italy's position was higher on almost all indices except for educational level of the labor force. Whether these differences make a difference in the way workers relate to various social systems will be explored in later chapters. The next section describes the communities in more detail and the factories selected for study.

Lansing

Lansing, Michigan, is located eighty-five miles northwest of Detroit. At the time of research, the city had a population of 110,000 with a metropolitan population of 300,000 and labor force of 110,000. Prior to the Civil War, the local economy was based on lumbering and services for the surrounding area. In 1860, although its population was only 10,000, the city already had 156 manufacturing firms, and by 1890 there were 215 (Darling, 1950:70). Thus, a small industrial base existed prior to the growth of the automobile industry. Before Ford began to build automobiles in Detroit in 1903, Ransom E. Olds founded the Olds Motor Vehicle Company in Lansing in 1897. Labor troubles allegedly led him to move the firm to Detroit, but in 1901 a machinists' strike in Detroit and the urgings of Lansing businessmen convinced Olds to build a factory in Lansing. The industry grew rapidly; by 1903, Olds employed 1,200 workers and produced 4,000 automobiles annually. For a time, Lansing was the center of the nation's infant auto industry. In 1908, Olds sold the factory to General Motors, but he built another to manufacture the REO (his initials). During the decade of 1900-1910 many other firms manufacturing rubber tires, auto bodies, wheels, and auto parts were attracted to the city, doubling the population (Niemeyer, 1963:42). In the next decade, the population doubled again as forges, steel fabrication, instruments, motor, and many other plants were established to furnish parts for the growing automobile industry. By the mid-1920's the pace of growth declined, and many small, locally owned plants were absorbed into large absentee-owned companies. The Great Depression slowed growth until World War II, when industry spurted again. After the war, the service sector grew rapidly with the expansion of government, education, and professional and business services. Lansing is now part of a vast industrial region stretching from Detroit to Lake Michigan and Chicago.

The Lansing metropolitan region has a more balanced economy than do most industrial areas. The labor force of the city is almost the size of its resident population because employees commute from as much as fifty miles away to work in the factories and in offices of state government, university, business, clerical, and professional services. Work organizations are large; in 1963, Lansing's 170 manufacturing plants averaged 108 workers (U.S. Bureau of the Census, 1966:23-28), and two-thirds of the wage earners worked in plants with over 500 employees. Government bureaus were typically large, and the State University employed 5,176 employees at the time of the research.

Lansing is a city of grey uniformity. Industries, typically concentrated near the city center, are surrounded by tidy working-class single dwellings. Except for three small slums which house blacks and recent Spanish-speaking migrants, the physical upkeep of the city is good. Surrounding the city are upper middle-class areas and suburbs which house professional workers and the managers of industry, government, and the university. Nearby new, large shopping centers thrive while the central business district languishes. Although the Lansing metropolitan area is growing faster than any other in the state, urban facilities are also being extended rapidly, giving the region an impression of prosperous and orderly growth.

A large metropolis may have represented the United States better than Lansing, but it is difficult to specify how the selection of a middle-sized city might bias the findings. Lansing's population had a smaller percentage of black and Appalachian residents than Detroit, for example, and the Lansing region was dominantly Republican in its politics. Most residents liked Lansing as a place to live and work. Typical of the labor force as a whole, one-half of OLDS employees resided outside the city in communities within a fifty-mile radius of the factory. This dispersed residential pattern of employees also occurred in the other three plants.

Turin

In 1961, when the field work for this study began, Italy celebrated its centennial as a united nation. Turin (Torino) played host to an international fair to commemorate this event, a fitting thing for the city which spearheaded Italian unification. In 1861, Turin was a city of 200,000 and the seat of the kingdom of Sardinia, ruled by the House of Savoy. Count Cavour was the chief architect of the unification, and the Savoys became Italy's royal family. A century later, the city had grown to over a million and was part of the Italian industrial heartland (Turin, Genoa, and Milan). Situated east of the French Alps along the rich agricultural Po Valley, Turin is the center of Italy's automobile and textile industries. Fourteen of FIAT's twenty plants, which together employed over 100,000 workers, were located in the city. In the late 1960's and early 1970's, FIAT was Europe's chief automobile producer as well as Italy's largest industrial combine.

The growth of Turin from a Roman town to one of Italy's major industrial centers is a fascinating story, which cannot be reviewed here. In 1400, Turin was a town of 4,000 inhabitants; by 1824 it had grown to about 110,000, the size of Lansing in 1960. At the time of Italy's unification in 1861, Turin had grown to 200,000. Fifty years later (1910), just before World War I, its population had doubled. And in another half-century (1960), it had tripled to over a million (Divisione Lavoro e Statistica, 1959:20).

At the time of unification (1861), local industry was largely at the handicraft stage. The largest factory was the arsenal, with its foundry and arms depot which employed 350 men (Fossati, 1951:226). But other industries were developing such as metal casting, clothing, tools, utensils, and machinery. Metal manufacturing was soon outpaced by textiles and clothing, which up to 1898 employed a majority of the labor force in the city and province. With the advent of Alpine electrial energy in 1899, the year FIAT was founded, Turin's economy, though experiencing recurrent depressions, grew steadily with the automobile industry (IRES, 1959:26-29).

FIAT (Fabbrica Italiana Automobili Torino) was founded in 1899, four years before Ford, on the initiative of a group of Turinese businessmen and aristocrats moved by both financial interests and their attraction to automobiling as a sport. Business interests soon dominated, and by 1908 the company also manufactured ball bearings, metals, coaches, diesel motors, and aircraft. At that time twenty automobile companies existed in the city (Fossati, 1951:221-259), yet all but two were liquidated by the Depression or absorbed into FIAT. The company produced

4,000 cars in 1912, a figure which OLDS had achieved nine years earlier. Tire, body (Ghia), accessories, and oil industries were attracted to Turin. Mountaineers from the unproductive Alps poured into the city during World War I to become part of the militant Turinese proletariat who became the avant-garde of the Italian labor movement prior to Mussolini.

At this time FIAT management decided to build a vertically integrated industry. Twenty plants were constructed including steel mills, aluminum foundries, tractor, truck, trolley, locomotive, diesel and electric engines. A period of growth during World War II was followed by a depression, but FIAT's fantastic growth during the decade 1950-1960 contributed greatly to Italy's economic miracle of industrial development. Though other large manufacturing plants were attracted to Turin, FIAT and mass-production industry do not completely dominate the city. The artisan tradition survives in the city's 25,000 shops; 6,000 mechanical workshops employ an average of two persons, while 7,500 clothing shops are one-person operations. The dual economy of a developing society can be seen in Turin, where the lone artisan still finds a place to live and work among industrial giants.

Prior to World War II, migrants to Turin came from Piemonte and the north, but after the war they came predominantly from the south. By 1962 about 300,000 southern Italians lived in the city. Most of them worked in personal services, construction, and other industries which needed unskilled labor, rather than in the large manufacturing plants. A severe housing shortage had developed but, although many apartments were overcrowded and unhealthy, few looked like slums. Working-class areas inside the city were well supplied with markets and institutional services. Neighborhoods were easily identified because they had official names and some had strong traditions. Some of the newer neighborhoods in the city's periphery had no local markets, and the city had yet to provide street paving, lights, sidewalks, sewers, parks, and schools.

Turin, a modern cosmopolitan community with a classically beautiful center, has a wide range of cultural facilities (museums, concert halls, galleries), educational institutions, smart shops, recreational facilities, and services. It is a well-planned city with wide boulevards, parks, playgrounds, and adequate governmental services. Most people live in relatively well-preserved five- or six-story apartment buildings. Despite the pervasive industrial smog, especially in winter, most residents, natives and immigrants alike, consider Turin to be a jewel of a city and feel fortunate to live there.

Córdoba

About 400 miles northwest of Buenos Aires, at the foot of the Córdoba Sierras, lies the old city of Córdoba, which in 1965 had a population of about 600,000. At the edge of vast pampas and astride the main trade routes to the north, the city was Argentina's largest in 1650. Today it is the nation's second city, but its economy is dominated by policies set in Buenos Aires. Traditionally, Córdoba has served as the political capital of the province, the service and educational center of a large rich hinterland, and a processing center for cereals and meats. The railroads, government offices, mills, agricultural storage facilities, and university constituted the traditional sources of employment. Factories making automobiles, trucks, motorcycles, airplanes, and farm equipment, and the small shops which supply these industries, are recent sources of employment. Since the manufacturing sector of the city prospered after 1950, Córdoba, unlike Lansing or Turin, is not a traditional industrial city. Although the city has been exposed to industrial and commercial influences for a long time, it accepts them hesitantly. The graceful colonial inner city is marred by stark apartments nearby. Antiquated noisy buses roar along main thoroughfares, but two blocks away pedestrians walk in the middle of the streets, forcing impatient motorists to edge their way through the crowds.

Urban-industrial culture boldly appeared in Argentina almost a century ago. "In the decade 1860-70, an accelerated process of economic growth and modernization was initiated which in a little more than 40 years completely transformed the social structure and economy of the country, at least its central region where two-thirds of the population live" (Germani, 1966:384). In 1895, over four-fifths of the proprietors and nearly two-thirds of the industrial workers were foreigners (Conde, 1965:74). This pattern of foreign ownership and employment persists to this day.

Early industrial development was mainly in the processing of agricultural goods. Foreign competition restrained the development of chemicals, oils, paper, textiles and other industries, holding the percentage of the labor force in manufacturing to twenty percent for the first half of the twentieth century. During World War II, industrial expansion occurred primarily in the light rather than the heavy industries. Yet by 1944 the net value of manufacturing output exceeded that of agriculture for the first time (Fillol, 1961:49). However, the proportion of the labor force employed in industry actually decreased between 1945 and 1947.

Peron's first five-year plan (1947-51) aimed to develop heavy industry and reduce the nation's reliance on agriculture. But surpluses in foreign exchange amassed during World War II quickly dissipated and employment in manufacturing declined between 1948 and 1954 (Fillol, 1961:53). Industry in Buenos Aires was simply unable to absorb the flood of European immigrants and rural migrants from the hinterland. Still, by 1955, local industry produced almost all the consumer goods used in the country (Scobie, 1964:227). Since that time, two important changes have occurred in the industrial sector: the dispersion of large industries to provincial cities and the development of heavy industries such as automobile manufacturing.

Milling flour, brewing, and food processing were Córdoba's first large industries. Yet most of the enterprises were small; in 1914, 2,836 establishments employed 19,081 persons. Many of the city's clothing shops were little more than cottage industries. Although data are not available to describe the changes in the city between the two wars, apparently the growth of the manufacturing sector lagged behind that of Buenos Aires. In 1927 DINFIA (Dirección Nacional Fabricaciones e Investigaciones Aeronauticas), a government agency, decided to build an airplane factory near the Córdoba army base and brought the city its first large factory. Soon after, the government attached a technical school to the plant to train skilled workers and give them work experience. The factory no longer builds airplanes, but it produces a light truck (the motors are built elsewhere), and the school still functions. Plants manufacturing parts were attracted to the city, and slowly a skilled industrial labor force was built up. Surplus factory space and a redundant labor force led the government after World War II to encourage FIAT and later Kaiser to locate in Córdoba. Whether the decisions to locate in the city resulted from a government policy to disperse industry or the attraction of lower labor costs, is unknown, but in 1955 Córdoba had three rather large automobile plants. FIAT and IKA employed over 5,000 each, and DINFIA more than 2,000.

A 1961 study revealed that the city's labor force had changed drastically. Chemicals, vehicles, and machinery factories employed 75 percent of the employees in manufacturing while the number employed in the traditional industries actually declined between 1958 and 1961 (Palmieri and Colome, 1964:38). Not only did IKA (Industrias Kaiser Argentina) become a major vehicle producer, turning out 50,000 cars in 1964, the company also became a supplier for other automobile manufacturers. Along with DINFIA, IKA established schools to train industrial workers for the needed jobs. It is difficult to predict whether the major incursion of large-scale manufacturing into Córdoba will be sustained. However, the city does have a large pool of well-educated, native-born (93 percent) industrial workers who earn above average incomes (Consejo Nacional, 1965). They constitute a self-conscious, unionized, politically alert stratum which demands steady employment whatever the market situation for industrial goods.

Córdoba is a city of openly visible contrasts. In its commercial core are modern offices, banks, restaurants, theaters, and government offices. Traditional and modern architecture exist side by side. Interspersed among the large and spacious buildings are the economically marginal shops and stores commonly found in small Latin American cities smart shops, and small art galleries. All structures, new and old alike, are so poorly cared for that they soon look dingy. Middle-class and some working-class areas near the city center have good markets, public utilities, and institutional services. But housing has not kept up with the population growth, especially since the first Peron era. Rents have been frozen but housing costs have soared, with the consequence that some working-class people pay low rent for superior quality housing while others pay high rent for inferior housing. Apartments and single dwellings recently built on the edge of the city vary widely in quality. Sewers, paved streets, lights, walks, as well as markets and institutional services have been extended unevenly. The absence of these facilities and services, more than the quality of the buildings, gives these areas the appearance of semi-slums. But genuine slum areas are not far away.

Paradoxes abound. The ancient university provides some of the best professional training in the nation, but it is also a training ground for radical intellectuals interested in working-class politics. A nearby army post shields a cadre of politically alert conservative generals who occasionally take control of the national government. A good symphony orchestra performs in an auditorium of faded Victorian elegance while city and state officials fight over who should provide schools for school-less children. Córdoba is a large traditional city in the industrial age, but it doesn't know where to go or how to get there.

Bombay

In a country where only 4 percent of the labor force works in factories like those found in the West, Bombay (population 4,500,000 in 1965) represents an industrial island in a rural sea. Since the city has long been considered a high-wage area, it has grown primarily through migration. Vaidyanathan (1971) reports that 85 percent of the work force was born outside the city and perhaps as many as one-quarter live there without a nuclear family. Males constitute 60 percent of the population of the city and 90 percent of the labor force. In such a city, industrial workers are considered an elite (Myers and Kannappan, 1970:133) because as many as 25 percent of the employable are casual laborers in the service sector and an estimated 30 percent are illiterate and unskilled.

A city of enormous contrasts, Bombay serves both as an overseas trade center and as a trade center for the western region of India. Although commerce is the main economic activity, the city also houses many large factories and thousands of small artisans' shops. Manufacturing dates back to the middle of the nineteenth century, when cotton and jute factories were built and a coal industry was developed to provide fuel for the newly established railroad system. The industrial growth of the country was confined to these four industries until the Tata iron and steel works were constructed in 1911. The absence of light industry prior to 1940 has been attributed to British policy, but after World War II and independence, many new industries appeared. Coal, textiles, and steel continued to grow moderately, but chemical, cement, and light industries increased twofold from 1946 to 1955 (Myers, 1958:16).

Motor vehicle production has been small; in 1965 eight plants produced only 69,500 units (Baranson, 1969:84). The state of Maharashtra, of which Bombay is the capital, is the most industrialized and urbanized in the nation, and it produced one-quarter of the industrial output of the country (Government of India, 1960:112). The plant I studied is Premier Automobiles Limited (PAL) which is located in Kurla (population 100,000), an administrative division of Greater Bombay situated on the southern tip of Salsatte Island. Only a creek separates Kurla from the urban area of Bombay.

Population density is very high, and housing is poor and overcrowded in the Bombay area. About two-fifths of PAL's employees lived in Kurla or immediately adjacent districts; the rest commute by bus or railroads, which run through the city (Rajagopolan, 1962:xv-xx). Workers live in apartments, chawls, compounds, or huts throughout Greater Bombay. Rarely do families have more than 150 square feet of living space under a roof (Sovani, 1966:80). Most residential areas have inadequate public utilities, markets, and institutional services, and complaints about them are commonplace. Although ubiquitous neighborhood associations supposedly handle water distribution and other problems, most residents regard them as inoperative but useful to local politicians. Contrary to popular belief, most working-class areas are not segregated by caste, and religious life is perfunctory (Wiebe, 1973). In a survey of factory workers, Gore (1970: 150-70) found that two-thirds considered themselves to be well off, three-quarters reported satisfactory social ties in the neighborhood, and about the same percentage reported being adjusted to life in Bombay.

As a cosmopolitan entrepôt Bombay has a wide range of cultural facilities, some fashionable apartment areas, and a bustling central shopping district. Workers seldom visit the central city and are not drawn into its commercial, political, and cultural life. They are in the city, but not of it.

The Four Cities

Do the national differences in technological level persist when the four communities are compared? The cities superficially resemble each other; each is the state capital; each, except Lansing, is the largest city of its state; each, except Lansing, is the most industrial city in its state. But the differences among the cities are dramatic. Lansing is a somewhat drab, middle-sized, Midwestern city with a balanced economy of manufacturing, government, and education. Turin is an industrial cosmopolitan center of a million which still retains some of its elegant splendor as the ancient capital of Savoy. Córdoba, with its half-million inhabitants and venerable history as the educational and trade center of Western Argentina, embraces modern industry with hesitation. Greater Bombay, a sprawling polyglot entrepôt of 4,500,000, has lost control of its development.

The four cities vary in the way they represent the industrial development of their nations. Lansing and Córdoba are quite representative, Turin is advanced for Italy, and Bombay is like a large commercial and industrial island in South Asia. Turin would strike a visitor as the most cosmopolitan city, followed by Córdoba, Bombay, and Lansing. The other three support a wider range of cultural services and institutions than does Lansing, but the large state university nearby gives Lansing access to cultural services. But since this study focuses on the extent that manual workers are exposed to an advanced industrial milieu, a good case can be made that workers in Lansing are the most exposed, followed by Turin, Córdoba, and Bombay.

Lansing's factories on the average are the largest and technologically most sophisticated. Lansing's population is most mobile and has easiest access to nearby cities and their educational, recreational, and other facilities. Lansing homes have the most mechanical and electrical appliances: television, radio, telephone, refrigerator, washing machine, and phonograph. Lansing's population is the most highly educated; it has the widest newspaper readership and is most exposed to channels of mass communication. Many recreational, fraternal, religious, and special interest organizations are available locally.

The greater exposure of Lansing's manual workers to urban-industrial culture is undoubtedly due to their higher standard of living. While all of Lansing's facilities are present in Bombay, for example, they are not available to ordinary manual workers. Clearly, Lansing's workers are the most exposed to industrial culture and Bombay's the least. Whether Turin's workers are more exposed than Córdoba's is harder to determine. Turin is a more industrialized, bureaucratized, cosmopolitan city than Córdoba; its services and facilities are better developed and more widely distributed; and its organizational life is richer. But these differences are tempered by the fact that the Argentinians have a high standard of living for their level of industrialization and they also are as well educated. If the Turinese have an advantage over the Córdobese, it is small.

The Four Factories

The automobile industry was selected for this study because its technology and organization are sufficiently standard in countries at different levels of industrial development to permit comparisons of worker behavior. Yet sufficient differences existed among the factories to justify a description of their physical, technical, and organizational environments.

OLDS

Oldsmobile Division of General Motors consists of a number of buildings which stretch almost from the center of Lansing to its periphery. As the plant grew over the years, it occupied adjacent residential areas and choice land along the Grand River. Most of the buildings are vast one-story sheds filled with machines packed closely together. The lighting and ventilation are moderately good, but the noise level is high. Visitors receive the impression that the plant is a large machine depot which turns out many cars with few workers.

OLDS pioneered in automated production. Engine blocks are automatically fashioned by large multi-purpose machines. Transfer machines move engines and other pieces from one production position to another. Overhead conveyor belts move supplies rapidly from one part of the plant to another. Semi-automatic machines which make parts are so closely spaced that one worker can monitor several with minimum movement. Sub-assembly lines articulate neatly, so that products flow swiftly to their destinations. The final assembly lines move rapidly as workers in pits and on the floor attach pieces with electrically powered equipment. Every forty seconds one car is driven off to the testing and parking areas. Although this bustling and well-ordered scene is typical of the industry, it always amazes visitors.

The experimental department which plans new models is shrouded in secrecy. A special pass, which only a few top officials have, is required for entry. Working conditions are unique; machines are widely spaced and ventilation and light are good.

White-coated engineers roam the department, scrutinize blueprints and talk to the skilled workers. The latter work unhurriedly; they often sit on their stools and monitor their machines as they go through set processes.

Employment in the automobile industry fluctuates considerably, but OLDS employment has been relatively stable for years. Workers exhibit a pattern of dual loyalty; they respect the company and have confidence in the union. Although a turbulent strike occurred in the late 1930's before the UAW (United Automobile Workers) was recognized by management, industrial relations have been calm for several years. The general collective bargaining agreement is worked out in Detroit, leaving only minor issues to be resolved locally. The UAW has been relatively successful in bargaining, so autoworkers are among the highest paid in manufacturing. A guaranteed annual wage, supplementary unemployment benefits, automatic cost-of-living wage adjustments, and prepaid medical insurance are some of the past bargaining achievements. In recent years, skilled workers have become disaffected with the union because they feel that officers favor the unskilled in wage negotiations. Although the skilled have a separate bargaining unit, they cannot veto a contract approved by the majority. Both the UAW and GM sponsor athletic and other activities to integrate the workers, but few participate. Workers largely ignore the rhetoric of America's chief speaker (GM) for private enterprise and labor's chief champion (UAW) of the welfare state.

FIAT

FIAT's position in Italy is comparable to General Motors' in the United States. FIAT, the country's largest enterprise, virtually monopolizes the production of automobiles, trucks, buses, planes, marine motors, locomotives, trains, refrigerators, and parts for all these products. As a vertically integrated organization, FIAT buys raw materials, makes them into finished products, and sells and services them through its own outlet. At the time of the study, the company employed 112,000 workers, 91,000 of which were factory workers. Of FIAT's 20 manufacturing establishments, the largest is the Mirafiore automobile works in Turin. This 420-acre modern (post-war) complex built at the edge of the city is half covered with workshops connected by seven miles of subways and fourteen miles of railroad tracks. Altogether

33,000 employees, including 3,000 office workers, work at Mirafiore. Some commute to work at least thirty miles by train, street cars, public buses, company-provided buses, and private cars.

The interior of the plant is a marvel of industrial planning. All sections are bright, clean, well-ventilated, and quiet by American standards. Automation is not as advanced as in OLDS, but it has been introduced. Men, machines, and materials move by overhead and surface conveyor belts. Ten thousand machine tools are neatly arranged in a large single-story building. Semi-automatic parts-making machines appear to be identical to those at OLDS. They are closely spaced, and workers monitor several of them simultaneously. Lines which assemble models of different size move at a somewhat slower pace than those at OLDS, turning out an average 3,600 cars a day. Since FIAT cars are smaller than OLDS', employees work closer to one another. The experimental department is secluded in a corner of the plant and a special pass is needed to enter it. In a light, well-ventilated, and quiet building, several autonomous units work on different parts of the new models. Interaction among the units and the workers in those units is minimal.

FIAT has the reputation of being a paternalistic organization which provides many services to employees but tries to influence their politics (cf. Minucci and Vertone, 1960:113-127). Wages are moderately high, the work week is 45 hours, overtime is required, and unemployment low. The company has a comprehensive social welfare system, one of the largest in the country. A health service (Mutua), a social work service, summer camps for children, day nurseries, and a veterans' retirement home are some of the services FIAT provides to supplement the national social security and welfare systems. Adult evening classes, sports facilities, hobby shops, and athletic programs appeal to some younger workers. The FIAT Central Training School, "G. Agnelli," located at the site of FIAT's first factory provides two to three years of technical training for promising youths to become skilled workers. Apprentices spend six months in the school preparing for work needed in various departments. Since 1955, large sums have been allocated to the "FIAT Housing Plan." The first 3,220 units, scattered in various neighborhoods of the city, were completed by 1962. These flats are rented at subsidized rates to those who show promise of becoming permanent employees. Many recent low-income migrants prized these apartments because housing was scarce.

After World War II and the reintroduction of independent labor unions, labor disputes multiplied, and workers occupied the factories with the intent of taking them over. After their eviction, FIAT undertook a successful campaign to reduce the influence of the communists and left-wing socialists. The task was simplified by splits within the labor movement. Four competing unions obtained representation in the enterprise: Confederazione Italiana Sindacati Lavoratori (CISL), a Catholic union; Unione Italiana del Lavoro (UIL), a social democratic union; Sindacato Italiano dell-Automobile (SIDA), an independent Catholic-oriented union; and Confederazione Generale Italiana dei Lavoratori (CGIL), a left-wing union. Union membership is voluntary in Italy, but all employees may vote for slates of the Commissione Interna (Internal Commission) offered by the unions. The Commissione is an elected body with proportional representation from all four unions. Since the contracts for all metalworkers are made at the national and provincial levels, the Commissione bargains for local adjustments to the contract, processes grievances, monitors the application of the contract, and prepares recommendations for future contracts. Only 15 percent of FIAT's employees were members of unions although almost all voted for representatives on the Commissione. That body was divided and weak, which made it possible for management to deal with each union separately, ignore the communist-dominated CGIL, and offer workers the kind of package it wanted (Carocci, 1960:225-335).

IKA (Industrias Kaiser Argentina)

Until the mid-fifties, Argentina relied heavily on foreign imports of automobiles. Peron, wishing to develop a local industry, encouraged Henry Kaiser to build a plant in Córdoba. An agreement was signed in January, 1955, which involved an investment of twenty million dollars; tools and equipment from Kaiser constituted one-third of the total, machinery from the state-owned auto plant represented one-fifth, and the remainder was a development loan from the Industrial Bank of the Republic. Kaiser sent a corps of managers to supervise plant construction and the organization of production. A few highly skilled workers were also sent to train Argentine employees. In April, 1955, plant construction began in a field five miles from the center of Córdoba. About a year later, the first Jeep was produced and by the end of 1956, 2,400 Jeeps and Ramblers had been built. Slowly, in shakedown operations, all but two or three American managers were replaced by Argentinians. In recent years, the factory has produced about 50,000 vehicles annually. Moreover, with the same sized labor force, it produced a larger proportion of automobile parts. In 1965, the company claimed that less than 10 percent of the parts was produced outside of Argentina. At that time, the company employed almost 10,000 workers, most of whom lived in Córdoba, but many commuted from a distance of thirty miles by public buses, private cars, and a fleet of old refurbished buses owned cooperatively by the workers.

IKA is a large cluster of well-designed, modern, single-story structures—well-lighted, well-ventilated, with such amenities as cafeterias, showers, recreational areas, and medical clinics. Though modern, the plant's mechanical equipment does not compare favorably with that of OLDS and FIAT. Automated machinery is almost nonexistent and the machines tend to be old American imports. Materials are handled with trucks rather than by conveyor belts and transfer machines. Since department boundaries are not clearly visible, work arrangements appear to be temporary. Parts-making departments most closely resemble those in OLDS and FIAT; semi-automatic lathes, presses, and other machinery, though somewhat older than the American and Italian models, produce at reasonably high rates. Testing, machine repair, and related departments which employ skilled workers contain smaller machines of somewhat older vintage than those in FIAT and OLDS, although they require similar skill to operate. Unfortunately, skilled workers were sometimes asked to make a limited number of identical pieces which could be made by special semi-automatic machines, but were not because the number of pieces required was too small. Such work requires little consultation with engineers, managers, or fellow workers. Units which housed skilled workers tended to be crowded, resembling production departments.

The final assembly lines deviate most conspicuously from those at OLDS and FIAT. Three parallel lines produce several vehicles: the large Jeep station wagon or pick-up, the small Jeep, and the large and intermediate models of the Rambler manufactured under an agreement with American Motors Corporation; several models of the Renault under French agreement; and the Borgward truck, with German cooperation. Though these lines are propelled mechanically, they moved slowly and spasmodically. The slow pace results from three conditions: each operator performs more operations on the average than in OLDS or FIAT, operators are shifted from one model to another too often, and the supply of parts are sometimes delayed. Due to inspection errors, missing parts, or other problems, vehicles which do not pass inspection are parked along the assembly lines to be repaired or completed later. Cluttered work space appears to be the normal condition. On a good day, a shift turns out a maximum of 300 vehicles.

IKA management originally hired a young labor force. The promise of high wages attracted a flood of applicants. Construction workers building the plant and young workers from the service sector were assigned easily learned production jobs. A few skilled workers were attracted from other industries. Since the skilled workers with industrial experience were in short supply, management carefully selected promising candidates for a program of occupational upgrading. In 1959, a special training department (capacitación) was instituted to give short courses in tool- and die-making. Three years later the IKA Instituto Tecnico was established to train technicians and toolmakers while providing a standard high-school education. Upon graduation, many trainees became IKA employees. Upgrading was also possible by going to night school, but more commonly unskilled workers selected the machines they would like to learn. After observing operators at work and after practicing on the machine, workers would request an examination for a higher skilled job.

On the whole, IKA had a good community reputation. A job with the company was highly desirable, and applications far exceeded vacancies. Industrial relations had been calm. The main labor union in the plant, SMATA (Sindicatos Mecanicos Automotores y Trabajadores Afines) was Peronista. Despite its political involvements, the local union retained considerable independence from Buenos Aires officials largely because the dues checkoff agreement with management made the local financially strong. Management's objective to concentrate on contract negotiations and grievances and avoid ideological disputes with the union was largely realized. The union had successfully negotiated automatic cost-of-living adjustments, paid vacations, illness benefits, participation in job evaluation studies, rate bargaining, and, unlike many Latin American unions, an effective grievance procedure. Most workers were favorable to both the union and management, except during periods of unemployment when, in traditional Argentine fashion, they demanded that management continue to employ all workers. However, strong political factions in the union continually challenged the faction in power because it overemphasized economic bargaining at the expense of political action.

PAL (Premier Automobiles Limited)

Although assembly plants have existed in India since the late 1920's, not until after World War II were complete automobiles manufactured. After independence, the Indian government stimulated automobile production by protecting local industry and closing foreign assembly plants. By 1965, six companies in the private sector produced about 50,000 units (Baranson, 1969:92). Bombay was a favorable location for the automobile industry because about one-third of the manufactured goods needed for automobiles is produced in the state of Maharashtra. A group of industrialists and businessmen incorporated PAL in June 1944. While the factory was being built on a ninety-acre marsh in Kurla, a group of young engineers were sent abroad to study automobile production. Arrangements were made with the Chrysler Corporation of the United States for PAL to build trucks, with FIAT of Italy to build passenger cars, and later with Fargo of Britain to build diesel trucks and buses. Assembly operations began in 1947; gradually more parts were manufactured so that by 1965 the plant made most of the parts. Since the first factory only assembled cars, most of the original employees were unskilled. When parts manufacture began, semiskilled workers were hired, and later skilled workers were added to make machine tools and maintain the machinery.

PAL's plants are flimsy compared to those of OLDS, FIAT, and IKA. But, like automobile factories everywhere, the buildings are large one-story sheds permitting maximum flexibility in the arrangement of machines. The tool room contains the most complex machines and employs some tool- and die-makers. Since new models are rare, the department makes and repairs machines to be used in production and occasionally produces needed spare parts.

The parts-manufacturing departments are small compared to the other plants, probably because of the low production schedule. Only forty vehicles a day are being produced, a figure far below plant capacity. Machines are crowded together and operators have only one or two semi-automatic lathes, drill presses, or grinders to monitor. Although set-up men prepare the machines for production, the operators make more adjustments than is customary for semiskilled workers. Skill differences between machine operators and assemblers are greater than in the other plants. No automated processes or mechanized conveyor systems are found in the plant. Motor blocks are placed on trolleys, pushed from one work station to another, and placed into position by manually operated pulleys.

PAL differs most from the other factories in the final assembly departments. Efficient assembly lines demand the synchronized movement of materials to workers as they perform operations on the moving chassis (cf. Walker and Guest, 1952:10). PAL lacks a mechanized system for conveying materials and its assembly lines are not mechanically propelled. Materials are brought to work stations on carts, and teams of two to six workers attach parts to the car or truck, which ride on a dolly. When operations are completed, the dolly is pushed to the next station, where another work-group performs its operations, and so on. Group leaders determine the work pace, and supervisors are responsible for meeting production quotas. Two parallel assembly lines, one for trucks and one for cars, stretch about 100 feet in a straight line. In these short lines, workers are very crowded, but they can readily see the entire line. Although the rapid pace so typical of assembly lines is absent in PAL, workers perform the same routine operations of assembly workers everywhere. PAL's technological and organizational structures broadly resemble the three other factories in the study, but the system results in low productivity and high product costs. Forty vehicles a day are produced at a unit cost 2.2 times that of making the same car in Italy (Baranson, 1969:33).

PAL's management attempts to project an image of a benevolent employer. In the Bombay labor market, wages in the automobile industry are relatively high and jobs were secure. The median annual wage of workers in the sample, in 1965, exclusive of fringe benefits, was 2,808 rupees compared to 1,866 for industrial workers in Maharashtra in 1962 (Sharma, 1974:12). The company had built some housing in a residential colony near the factory and an elementary school for employees' children. However, only a small proportion of the employees, mostly white-collar workers, could take advantage of these facilities and those of the local colony club, which sponsored hobbies, sports, drama, and recreational events. The company provided limited medical and health services, and bus transportation to rail points, easing the journey to and from work. About two-thirds of the employees believed that PAL was a better employer than others in the Bombay area.

When the factory began operations, a local of the Gandhian INTUC (Indian National Trade Union Congress) represented the workers, but after a long and bitter strike in 1958 lasting 110 days, management recognized the socialist EMS (Engineering Mazdoor Sabha). Since that time, labor-management relations have been calm. Although union membership was optional, approximately 80 percent of the workers belonged to one of the two local unions. Contract negotiations occurred every two or three years on the state level. Union officials were not deeply involved in the daily issues of the plant, but they did seek to raise the wages of individual workers. Only a few were involved in union affairs; like typical Indian workers, they were apathetic (cf. Sheth, 1968:163-67).

Table 2.3 presents data on the characteristics of the four factories. They differ in the complexity of their technologies and in the industrial experience of their personnel. OLDS is an old enterprise and has the most sophisticated technology, the most orderly growth, the most professionalized and experienced management, and the most experienced labor force. FIAT is the oldest plant, and it too has an advanced technology. Although its experience with large-scale production is more recent, it has a capable core of managers and skilled workers to handle problems of growth. IKA is the newest plant, but its technology is ten years behind that of the United States. Though both management and workers were trained rapidly, they function independently and post creditable production achievements. Although PAL is eleven years older than IKA and has the same-sized work force, its production record is dismal. Its machines and dies are outmoded and worn out and management has not learned how to use labor efficiently. In sum, although the four plants vary in technological sophistication and the managements differ in their organizational experiences, workers in parallel departments perform similar jobs. I shall now describe how the samples were chosen.

The Samples

Selecting samples of workers in the four plants had to take into account differences in plant technology and skill composition. The Mirafiore complex in Turin was the most integrated because it fabricated almost all of the parts of the automobile. OLDS shipped in bodies from the Lansing Fisher Body factory and sub-contracted for many parts. IKA stamped bodies from dies made in the United States and sub-contracted for a few parts. PAL fab-

Table 2.3 Characteristics of Four Automobile Plants

Characteristics	*OLDS*	*FIAT*	*IKA*	*PAL*
Date of founding	1902	1899	1955	1946
Rate of expansion	steady growth	gradual, rapid after W. W. II	rapid, recent curtailment	slow, recent expansion
Number of employees	12,000	33,000	11,000	8,200
Highest technology	high automation	simple automation	mechanical	simple mechanical
Vehicles per year	505,555	795,504	50,042	8,200
Percent indigenous parts	100	98	90	85
Experience of labor force	continually expanding core of skilled	experienced skilled and many new recruits	new core and new recruits	old unskilled core and new skilled recruits
Foreign management	none	none	few and diminishing	occasional consultants
Earnings ratio of skilled to unskilled	1.5	1.5	1.6	1.5
Mean annual income U.S. $ (1966)	8,212	2,542	2,173	591

ricated almost everything, but imported electrical components from FIAT. I had overestimated the uniformity in technology of the four factories: OLDS and FIAT were advanced exemplars; IKA was a decade behind the two; and PAL, a generation. Yet all four factories unmistakably were more than assembly plants. IKA and PAL managements insisted that they were true automobile manufacturers because they fabricated more of their parts than most American firms do. All four companies had departments made up mostly of skilled workers who either worked on new models or made tools, dies, and machines for the production departments. All companies had departments exclusively devoted to making automotive parts, motors, and accessories. Semiskilled workers in these departments operated pre-set semi-automatic machines. Finally, all factories contained assembly departments, including the final assembly, whose workers were unskilled.

Most departments were occupationally homogeneous; over 90 percent of the workers had similar skills. This fact was determined by noting the skill required to operate the machines, and not by accepting management's or the worker's definition of skill. In PAL, many long-tenured unskilled and semiskilled employees were classified and paid as skilled; in IKA, assembly-line workers who worked on more than one type of vehicle were classified as semiskilled. In FIAT, some employees were classified at one skill level but worked at another. In short, managers, workers, and researchers differed in their definitions of skill. Clearly, the definitions of skill represented temporary agreements among management, union, and the worker arrived at through bargaining. The more advanced the technology, the more the three groups agreed on the classification of jobs and the more universalistic was the application of the classification.

Departments directly involved in automobile manufacturing were classified according to their dominant skill level, to determine the skill composition of the factory. I excluded departments dominated by managers, engineers, and clerks, as well as those engaged in food preparation, sanitation, construction, transportation, salvage, storage, handling materials, and the like. One or more departments at each skill level was randomly selected as the pool from which the sample would be drawn. This procedure resulted in selecting the experimental department or the tool room for the pool of skilled workers, one of the final assembly departments for unskilled workers, and one of the parts production departments for the semiskilled workers. Lists of employees in selected departments were obtained from management, except for OLDS, where they were obtained from the union. Quotas were established for each department, and a random sample was drawn to fill the quotas. In FIAT, we wanted to make certain that the sample also adequately represented the four main labor unions. On the basis of departmental data on the last union election, we computed average optimum allocations for the four selected departments and sampled accordingly. Workers who had been employed less than one year were eliminated from the samples. Letters were sent to workers' homes describing the study and informing them that both management and the unions had approved it. Except for PAL, where workers were interviewed in the factory, all were interviewed in their homes (see Appendix B).

Table 2.4 presents the skill composition of the samples and the operations performed by the workers. Since the factories vary in the percentage of the automobile they manufacture, the distribution of skills and operations vary. The main differences are that the newer plants (IKA and PAL) employ relatively more unskilled workers, the older ones (OLDS and FIAT) more semiskilled workers, and PAL more skilled workers. The plants dif-

TABLE 2.4 SKILL LEVEL AND WORK OPERATIONS OF REPRESENTATIVE SAMPLES OF AUTOWORKERS IN FOUR PLANTS (PERCENTS)

	OLDS	*FIAT*	*IKA*	*PAL*
Skill Level				
Skilled	19	18	16	25
Semiskilled	53	51	35	39
Unskilled	28	30	48	36
Total	100	99	99	100
Operations				
Craft	12	18	10	30
Test, inspection, repair (TIR)	34	16	26	8
Machining	27	33	31	31
Assembling	27	32	33	31
Total	100	99	100	100
N	(249)	(306)	(275)	(263)

fer most in the proportions engaged in test, inspection or repair (TIR), and in craft operations. OLDS employs more workers in quality control; PAL more in craft operations. OLDS is most highly mechanized and needs more workers to monitor the quality of work turned out by automatic and semi-automatic machines. PAL, on the other hand, needs more craft workers because it does not have as many special automatic machines to make specific parts. The high percentage of IKA workers in TIR reflects not technological complexity but the fact that the factory produced a great variety of parts for different types of vehicles (Jeep, Renault, Rambler, Borgward).

The purpose of this detailed analysis of the labor forces of the factories is to demonstrate that autoworkers are much more diversified in their skills and functions than most people seem to think. In the following chapter I shall trace the social origins of workers in the various skill levels and demonstrate how they got to be where they are.

3

Stratal Origins and Destinations

Recruiting labor in industrial societies is a stratification phenomenon. The occupations of a factory are stratified according to their pay and attractiveness. The higher-paying jobs require expensive training and attract recruits who can afford the training, thus perpetuating the stratification system. As changing technologies alter the demand for particular occupations and as the supply of qualified workers fluctuates, wage relationships change, producing changes in the stratification system. Clearly, the factors which affect the labor market and the stratification system change as societies industrialize. Although a given industry (automobile) needs workers with the same occupational skills in countries of high and low industrialization, the workers in different countries may be recruited from different social strata. That is, the stratification dynamics of labor-force recruitment varies according to the extent of the society's industrialization.

Automobile manufacturing companies most need low-skilled assembly workers and semiskilled machine operators, who together constitute about 60 percent of the personnel. About 10 to 15 percent of the labor force is occupied in inspection, testing, and repair services and another 25 percent in skilled craft work. The supply of qualified employees for these occupations varies from one country to another. Where educational levels are high, almost all adults can perform most of the unskilled and semiskilled jobs without special training. Everybody knows something about machines, how to follow simple instructions, and how to comply with organizational requirements. When jobs are plentiful, people with above-average education avoid jobs in the automobile industry and seek work which pays more and has more prestige. The auto industry is then forced to hire "marginal" workers with low education and little industrial experience: e.g., rural migrants, foreign immigrants, or inexperienced young workers. Skilled workers, on the other hand, require long and specialized training. Typically they are in short supply and so well organized that they command high wages. Employers rarely train semiskilled workers for skilled jobs because it is too costly. Compared to the unskilled, craft workers tend to be older, more experienced, more highly educated, and more urban in origin. In the auto industry their wages average 50 percent higher than those of unskilled workers. The differences in social background and economic power of the skilled and the unskilled are dramatic. In the American context, for example, the unskilled assembler may be an Appalachian migrant and the skilled, an aristocrat of labor; in stratification terms, the range is from lower- to middle-class.

In societies where the general level of education is low, not everyone can perform unskilled or semiskilled industrial jobs. While managers may overestimate the amount of education and experience needed for unskilled factory jobs, not everyone knows how a machine works, how to follow instructions, and how to adapt to factory regimen. New employees may be trained for as long as two weeks for assembly-line jobs, in contrast to a day or two in the United States. Moreover, since unemployment is often so high in developing societies, many workers seek high-paying, steady, industrial jobs. This situation enables employers so to raise educational requirements that they can employ an educational elite, by local standards.

Requirements for semiskilled jobs are set even higher: e.g., the ability to use a gauge, follow written instructions, start and stop machines, and recognize machines' malfunctioning. Such skills can be quickly taught to literate unskilled workers in industrial societies. In less developed societies, highly educated urbanites are recruited into these jobs. Typically, their fathers have worked in the service sector either as shopkeepers or as casual laborers. These highly educated urban semiskilled workers represent a valuable labor pool which can profitably be trained for skilled jobs. In fact, such upgraded employees constitute the bulk of the skilled factory labor in countries of low industrialization. Obviously they are not as skilled as craftsmen in American factories. Such workers would be regarded almost as engineers in underdeveloped societies.

A highly industrial country like the United States obviously recruits factory workers in different ways from those of an underdeveloped society like India. But nations which fall in between these extremes develop unusual patterns. For example,

Argentina has a highly urbanized and highly educated labor force which can meet the personnel demands of a more industrial economy. For generations, Argentina has had some industry and a core of skilled artisans capable of performing very complex jobs. In short, Argentina has most of the characteristics of an industrial society—except a lot of industry. More skilled workers than jobs are available at times in some localities, so both Argentina and Italy have exported craft workers to other countries. Not uncommonly in this situation, the skills of workers are poorly utilized and craftsmen are asked to do semiskilled work. Frustrated in their aspirations, they press for radical changes in the society (Soares, 1966; Iutaka, 1963).

Although the automobile industry recruits a stratified labor force everywhere, the details differ in each country. The more industrial the nation, the less the unskilled and semiskilled differ in their stratal origin and training. Thus, the most important cleavage in the labor force of industrial countries is between semiskilled and skilled workers, whereas in the less industrial nations, the cleavage between the unskilled and semiskilled remains wide. The range of skills among factory workers is widest in highly industrial societies because craftsmen there perform more complex operations than craftsmen in less industrial societies, whereas unskilled labor is the same everywhere. Finally, the position of automobile workers in the stratification system differs in various countries. In the industrial ones, the unskilled constitute a low stratum, while the skilled are in the middle. In nonindustrial countries, the unskilled and the skilled are near or in the middle strata in terms of their education, present income, and life style because below them is a large class of marginal, casual, service laborers. Workers recognized this situation; only 27 percent of OLDS workers identified themselves as middle-class, as opposed to 61 percent of IKA's and 85 percent of PAL's. Whether differences in stratal position and class identification have political consequences cannot be determined *a priori*, but we shall examine the question later.

Social Origins

To what extent do data on the four plants conform to the above theoretical speculations on where autoworkers fit in the societal stratification system? Their social origins may provide a clue.

Considerable research supports the view that rural migrants to the city occupy jobs at the bottom of the occupational ladder because they lack specialized training for industrial jobs, have little knowledge of the labor market, and low job aspirations (Bendix and Lipset, 1952; Sewell, Haller, and Ohlendorf, 1970). Urban dwellers get more training, are more aware of job opportunities, and have greater access to them. However, highly educated job applicants from rural areas find better jobs than poorly educated urban applicants and highly educated sons of laborers find better jobs than poorly educated sons of white-collar workers (Jencks *et al.*, 1972:180-185). These generalizations should hold for all societies, regardless of their industrial development.

Yet these generalizations must be applied cautiously in comparative studies because attributes of industrial recruits must be evaluated in the context of local norms. Thus workers with the same education in different countries may be considered well educated or poorly educated depending upon the national average. But even national averages vary by age cohorts: older workers may have better jobs than younger workers, even though the latter may be better educated.

OLDS and FIAT respondents were older and had been employed longer than IKA's and PAL's employees. Even though younger workers may still be occupationally mobile, most mobility is achieved in the first ten years of work-life. Since the average employee in all factories worked fifteen years or more, most had probably become occupationally stable. This is suggested by the data on number of jobs held; regardless of differences in median ages, respondents in all plants averaged between three and four jobs.

Data in Table 3.1 tend to support my theoretical speculations. In the more industrial countries (United States and Italy), a larger percent of the workers were born in rural areas or small towns, while in the other countries, more were born in cities. Moreover, more OLDS workers' fathers were farmers, while more fathers of IKA and PAL workers had worked primarily in the business or service sectors. While one-quarter of the fathers in all countries except India were factory workers, more of the fathers of OLDS and FIAT employees were artisans who found the transition into manufacturing easy. (See Table A.1, p. 276.)

Considering the proportion of the national labor force in agri-

TABLE 3.1 SOCIOECONOMIC BACKGROUNDS AND DEMOGRAPHIC CHARACTERISTICS OF FOUR SAMPLES OF WORKERS

Characteristics	*OLDS*	*FIAT*	*IKA*	*PAL*
Birthplace: rural or small town (percent)	68	66	36	40
Father's sector (percent)				
Agriculture	40[a]	34	21	35
Manufacturing	30	24	26	10
Craft-artisan	19	21	7	9
Business and service	11	21	46	46
Totals	100	100	100	100
Years of education (median)	9	5	7	6
Sector background (percent)				
Manufacturing	92	59	62	47
Business and service	3	12	32	28
Agriculture	6	19	2	14
No previous employment	9	10	4	11
Totals	100	100	100	100
Number of years in the labor force (mean)	25	22	15	15
Number of occupations ever held (mean)	3.8	3.3	3.7	3.9
Mean age	43	35	30	32
Age at first full-time job (mean)	17	13	15	17
Tenure in factory (mean years)	13	9	5	8
Married (percent)	94	76	71	81
Number of children (mean)	2.80	1.07	1.55	2.21
Wife employed (percent)	29	26	15	5
N	(249)	(306)	(275)	(263)

[a] Includes 5 percent with half of their careers in manufacturing.

culture at the time when workers in the study were born, it is clear that the Americans were disproportionately recruited from agricultural backgrounds. Thus, in 1920, the mean year OLDS workers were born, 40 percent had fathers who were farmers, compared to only 15 percent of the nation's labor force. In contrast, while over 85 percent of Indians worked in agriculture in 1935, the mean year PAL workers were born, only 35 percent had fathers who were farmers. Argentina is similar to India as a case where industrial recruits came from the nonagricultural sector, but the Italian situation is reversed and close to the American. Briefly, the more industrial the society, the more industrial recruits had agricultural backgrounds and the lower they were in socioeconomic status.

With respect to education, OLDS workers had a median of nine years, the same as their age cohort's, while PAL's averaged six years, a level attained by only a small minority in India, which was over 70 percent illiterate in 1960. Italian and Argentinian workers were also more educated than their age cohorts. When the education of the samples is compared to that of the adult males of the cities in which they resided, OLDS' employees were below average, FIAT's were average, IKA's, above average, and PAL's were vastly above the average.[1] National differences also appeared in apprentice training. Less than 10 percent of OLDS workers had such training, in contrast to 20 percent of IKA's and three-tenths of FIAT's. Clearly, all workers except the American were an educational elite; the lower the industrial development of the nation, the more highly educated they were by local standards.

Finally, I examined the sector work experience (agriculture, manufacturing, business or service) of the four samples. All occupations that workers had ever held were coded for their appropriate sectors and then each career was categorized for the sector in which the individual had worked the longest (see Table 3.1). Less than one-fifth of all workers had had significant agricultural employment; the majority had worked mostly in manufacturing. The percentage in each sample that had dominant manufacturing work experience increased with the extent of national industrialization and not according to the fathers' work sector or rural-urban background. Thus, while two-fifths of OLDS employees had fathers who were farmers, four-fifths of the sons had worked mostly in manufacturing. In contrast, a third of the PAL sample had farmer fathers, but barely half of the sons had dominant work experience in manufacturing. Twice as many persons in OLDS and FIAT than in IKA and PAL had worked only in manufacturing.

Briefly, the more industrial the country, the more industry

[1] Median years of education for adult males: Lansing, 11.5; Turin, 4.8; Córdoba, 5.9; Bombay, 3.0. In every instance except that of the United States, the city average was higher than that for adult males in the state or province.

rapidly absorbs employees with agricultural backgrounds. Two explanations for this are possible. First, rural-reared factory recruits in industrial societies have sufficiently absorbed the endemic urban-industrial values to accommodate easily to factory work, but urban socialization is needed for people reared in nonindustrial societies (see Chapter 4). Second, employers in industrial societies have no choice but to hire the most marginal applicants, i.e., rural migrants with low educational achievement who flood the cities. The second explanation is more likely because workers with widely different backgrounds all build automobiles; FIAT's employees have less formal education than IKA's and PAL's, but they are more productive. Managers probably do not know how much education and urban experience workers need to be productive. In nonindustrial societies, managers simply set educational requirements for employment as high as possible.

Social Origins and Occupational Mobility

Autoworkers in industrial countries are recruited from lower social strata than are workers in nonindustrial societies, but do career and generational occupational mobility also vary with industrial development? The same variables used to analyze social origins were used to examine occupational mobility: community of origin, socioeconomic status of the family, and educational achievement. The industrialism hypothesis is that mobility should be determined in the same way in all societies: more of the unskilled should have rural origins, little education, and fathers with little urban-industrial experience. Moreover, the skilled should be recruited from urban-industrial backgrounds and from higher-status families who have provided their sons with above-average education. The development hypothesis considers differences in stratification systems of nations. Less developed societies tend to be more rigidly stratified and less universalistic than industrial societies, thus restricting educational opportunities to a larger extent. Consequently, skilled jobs go to a veritable educational elite.

In the comparative study of occupational mobility, one has to make a dangerous but unavoidable assumption: that the occupational hierarchy is similar in all countries (Miller, 1960:10-13). While some evidence supports this generalization (Hodge, Treiman, and Rossi, 1966), Haller, Holsinger, and Saraiva (1972) suggest that the researcher exercise caution especially when interpreting data from nations outside the Euro-American cultural system. No one has solved the vexing problem of ranking agricultural work in the occupational hierarchy. For this study, occupations are ranked from low to high as follows: farming, unskilled labor in service or manufacturing, semiskilled, skilled, office or sales, managerial or official, and professional work.

Two measures of occupational mobility were constructed, career and generational. To measure career mobility, all job changes were coded for the number of occupational levels moved up or down from the previous jobs. Careers were classified as evidencing upward, downward, or no mobility on the basis of the algebraic sum of all job moves. Occupational mobility may be steadily up or down or it may fluctuate.[2] To simplify the analysis, careers were summarized as exhibiting up, down, or no movement, ignoring fluctuations and the extent of mobility. Employees who have held only one job cannot experience mobility; this applied for less than 5 percent of the cases. To measure generational mobility, I compared the son's occupational level to his father's.

Career mobility patterns were similar in the four countries. About half of the workers experienced upward mobility; about a third, no mobility; and a tenth, downward mobility (see Table 3.2). PAL's employees were somewhat more upwardly mobile than those in other countries and IKA's were the least mobile. Table 3.3 provides information only for those who had experi-

TABLE 3.2 OCCUPATIONAL MOBILITY PATTERNS (PERCENTS)

	Companies				
Mobility Pattern	*OLDS*	*FIAT*	*IKA*	*PAL*	*Total*
Upward	51	48	43	56	49
No mobility	30	37	44	28	35
Downward	11	10	9	14	11
One job	7	5	4	3	5
Totals	99	100	100	101	100
N	(249)	(306)	(275)	(262)	(1092)

[2] Fluctuating careers were: OLDS, 35 percent; FIAT, 25 percent; IKA, 33 percent; PAL, 38 percent.

TABLE 3.3 OCCUPATIONAL MOBILITY FOR FIRST AND SUBSEQUENT JOB MOVES[a]

Job Moves	*Horizontal*	*Up*	*Down*	*Total*
OLDS				
First move	42	43	16	101
Subsequent moves (2-7)	54	29	16	100
All moves	50	33	17	100
FIAT				
First move	45	45	10	100
Subsequent moves (2-5)	53	27	20	100
All moves	50	35	15	100
IKA				
First move	54	36	10	100
Subsequent moves (2-5)	50	28	22	100
All moves	52	30	18	100
PAL				
First move	44	43	13	100
Subsequent moves (2-6)	58	25	17	100
All moves	54	30	16	100

[a] These data exclude workers who held only one job. An analysis was made of the directionality of every job move. The distribution was almost exactly the same for every move after the first, in each nation.

enced mobility. Although national patterns were distinctive, FIAT's and PAL's were most nearly alike. An equal proportion of the first moves were upward and horizontal at OLDS, but more of the subsequent moves were horizontal. FIAT's and PAL's patterns were similar to OLDS, except for slightly more downward mobility on moves subsequent to the first. Finally, for IKA, the proportion of horizontal moves remained almost the same on first and subsequent moves, but the extent of downward mobility increased after the first move. Thus, except for the United States, more downward mobility occurred after the first move. This suggests that workers may move into higher-paying factory jobs after holding low paying but higher status white-collar or skilled jobs. That is what actually occurred: over 25 percent of IKA and PAL employees had previously held white-collar jobs, compared to 15 percent for OLDS. For FIAT, 26 percent had been skilled workers or apprentices prior to working for the enterprise, compared to 15 percent for workers in the other companies.

Three social-origin variables were related to career mobility: community of socialization (where the person lived at 10-20 years of age), occupation of the father, and education of the son. These variables were all significantly and moderately associated with each other according to the chi-square test in all four countries.[3] Occupation of father and community of socialization were most highly correlated, followed by occupation of father and education of respondent, and finally community of socialization and education of respondent. The sizes of the correlations were small enough to suggest that each had an independent effect on mobility. Since all mobility data is cumbersome to present, Table 3.4 presents data only for the upwardly mobile and the nonmobile.

Only in Italy was community of socialization associated with career mobility; the urban socialized (mostly second-generation Turinese) experienced the least upward and the most downward mobility. In the other countries, those socialized in rural communities experienced most upward mobility. Upward mobility was related to level of the father's occupation in three countries. As predicted, the less industrial the country, the more were the father's occupation and the son's mobility associated: the association was random in the United States, statistically significant in Italy and Argentina, and highly significant in India. Most of this mobility reflects the upward mobility of farmers' sons. To some degree, it is an artifact of the scoring procedure because sons of farmers had nowhere to go but up. But some climbed more than one level. In Italy and Argentina, more than in the United States, factory workers' sons inherited their fathers' occupations, and in India skilled workers' sons exhibited more upward mobility than did similar workers in other countries. Finally, education was more highly associated with career mobility in all countries than were the other variables, but not in the expected way. In the United States and Italy, workers with low levels of education were more upwardly mobile than were the highly educated. This reflects the fact that older workers with less education than young workers held more skilled jobs and that

[3] Throughout this book, probabilities of chi-square at or below 5 percent are reported as significant and those from 5 to 10 percent as trends. The corrected coefficients of contingency ($\overline{C}$) ranged from .228 to .442. In this analysis OLDS and IKA samples respectively had 35 and 26 skilled workers added to the samples because there were so few in the original samples. Of course, tests of significance are inappropriate in such instances.

TABLE 3.4 OCCUPATIONAL MOBILITY BY COMMUNITY OF SOCIALIZATION, FATHER'S OCCUPATION, AND EDUCATION OF SON (PERCENTS)[a]

	OLDS		*FIAT*		*IKA*		*PAL*	
	Occupational Mobility Patterns							
	None	*Up*	*None*	*Up*	*None*	*Up*	*None*	*Up*
Community of Socialization								
Rural	24	56	38	51	40	46	30	58
Urban	29	51	44	22	33	47	—	—
Metropolitan	27	55	37	46	46	44	25	53
Totals	29	54	38	48	42	45	27	55
	p = N.S.		p = <.05 $\bar{C}$ = .252		N.S.		N.S.	
Father's Occupation								
Farmer	29	61	32	59	42	50	24	65
Factory	30	48	47	41	47	46	34	54
White-collar	28	52	36	40	34	43	28	44
Skilled	26	58	37	48	38	43	36	50
Totals	29	55	37	48	42	46	28	56
	p = N.S.		p = <.05 $\bar{C}$ = .275		p = <.01 $\bar{C}$ = .309		p = <.001 $\bar{C}$ = .481	
Education of Son								
Low	26	64	40	54	47	44	37	57
Middle	—	—	41	39	43	47	25	56
High	30	50	24	41	32	46	26	55
Totals	28	55	38	48	42	45	28	56
	p = <.05 $\bar{C}$ = .220		p = <.001 $\bar{C}$ = .384		p = <.001 $\bar{C}$ = .383		p = <.05 $\bar{C}$ = .291	

[a] Reliable data for PAL existed only for two categories of community of socialization and the range of education in OLDS was too small for three categories. Probabilities (p) of χ^2's and $\bar{C}$'s are based on original tables which include data on downward mobility and holding only one job.

old employees had been upgraded during periods of labor shortage. In Argentina and India the reverse condition prevailed; workers with low education remained nonmobile throughout their work lives.

In summary, community of socialization was not important for career mobility except for Italy, where the rural were more upwardly mobile than the urban. Whether this resulted from FIAT's policy, rumored as favoring rural workers, is difficult to tell, but in all countries, farmers' sons were the most upwardly mobile. As hypothesized for the less industrial countries, upward mobility was more tied to level of fathers' occupation and the amount of education given sons.

Intergenerational Mobility

People can be upwardly mobile in their careers and generationally downwardly mobile, and vice versa. I calculated two measures of generational mobility: one which classified farmers as unskilled and the other as below unskilled. Since the two measures yielded similar results, Table 3.5 provides data ranking

TABLE 3.5 INTERGENERATIONAL OCCUPATIONAL MOBILITY (PERCENTS)

Patterns	*OLDS*	*FIAT*	*IKA*	*PAL*
Son below father	26	30	36	38
Son same as father	29	32	34	25
Son above father	44	37	30	37
Totals	99	100	100	100

farmers the same as urban unskilled workers. Obviously the amount of generational mobility in a population is affected by the occupational composition of their fathers. The average occupational rank of the Indian and Argentinian fathers was identically high (4.0), followed by the United States (3.2), and Italy, the lowest (2.9) (see Table A.1 in Appendix). In the most industrial nations (United States and Italy) where the mean occupational ranks of fathers was lowest, sons exhibited the most upward mobility; in nations of moderate industrial development (Italy and Argentina), sons showed the most nonmobility; and in the least industrial nations (Argentina and India), sons exhibited most downward mobility. These trends only partly reflected the occupational composition of the fathers. Though fathers of workers in OLDS and FIAT had similar occupations, sons in FIAT experienced less vertical mobility. Fathers of FIAT and IKA workers differed greatly in occupational status, but their sons were similarly nonmobile. However, the greater downward mobility among IKA and PAL employees reflected the higher occupational status of their white-collar fathers (32 and 37 percent respectively compared to 10 percent for OLDS and FIAT). Finally, although age and seniority were related, they were not associated with mobility patterns anywhere.

These trends must be interpreted with caution because they are based on American conceptions of occupational prestige. In underdeveloped societies, where uncertainties in the economy make many businessmen little more than peddlers and clerks little more than custodians, acquiring an automobile job may represent upward status- and income-mobility. But genuine downward mobility appeared in all the samples even when the criteria for mobility were made more rigorous. Thus, when the occupations of fathers who were manual workers were compared to their sons', a larger percent of the fathers in all countries except India were skilled workers.

Career and generational mobility were moderately associated in all countries. Workers who were more upwardly mobile generationally were also more upwardly mobile in their own careers (see Table 3.6). Those who experienced no career mobility, were

Table 3.6 Generational and Career Mobility of Autoworkers (percents)

	OLDS		*FIAT*		*IKA*		*PAL*	
	Career Mobility							
Intergenerational Mobility[a]	*Up*	*None*	*Up*	*None*	*Up*	*None*	*Up*	*None*
Sons below father	43	36	37	44	34	48	47	31
Sons same as father	54	22	49	36	44	42	43	48
Sons above father	62	27	55	33	57	36	68	21
Totals	55	28	48	38	45	42	50	28

[a] Based on a scheme that classified farmers as below unskilled.

more downwardly mobile generationally, especially in Italy and Argentina. Finally, the less industrial the country, the more were career and generational immobility associated, supporting the development hypothesis that less industrial countries are more rigidly stratified (Cutright, 1968:405-11).

Origins and Occupational Mobility by Skill

For full understanding of mobility, the point of origin (father's occupation, community of socialization, and education) and the point of destination (unskilled, semiskilled, and skilled labor) must be considered together. Thus, while sons of farmers may be the most upwardly mobile, they may not occupy the highest occupations in the factory. This section examines the social origins of workers at each skill level as well as their occupational careers.

In all nations except Argentina, more of the unskilled and semiskilled workers had rural origins, and more of the skilled were reared in metropolitan centers. In fact, more of the skilled were born in the city in which the factory was located, except for Córdoba, which imported more skilled workers from other cities. With the exception of Argentina, the higher the skill of the worker, the more likely that he was socialized in a large city.

As might be expected, sons of farmers dominated unskilled work in every factory, but sons whose fathers were nonskilled factory or manual workers had better jobs than did sons of farmers in FIAT and IKA. In OLDS and FIAT, sons of craftsmen were disproportionately favored in skilled work. Finally, in all countries except the United States, sons of white-collar workers tended to be more highly represented in semiskilled and skilled work. In short, a working-class background, and especially a background of skilled work, favored upward mobility in the industrially advanced countries, and a white-collar background favored upward career mobility in the less advanced countries (see Table 3.7). Education was the variable most strongly associated with skill attainment; everywhere, the higher the worker's education, the higher his skill. As predicted, this trend was weakest in the United States, where education is most widespread, and strongest in Argentina and India, where educational opportunities are more limited.

In summary, in all countries workers with backgrounds of opportunity moved into the best jobs. The three indicators of social status operated similarly everywhere, but education was more strongly associated with occupational destination than were community of socialization and occupation of father. Although national differences were small as the industrialism hypothesis predicts, the data also support the development hypothesis that a background of privilege is more important for upward mobility in less developed countries.

Trajectories for Skill Levels

Sociologists who have analyzed hundreds of national surveys (Hamilton, 1967; Glenn and Alston, 1968) have found that, though skilled workers differ from both white-collar and other

TABLE 3.7 SOCIAL BACKGROUND AND OCCUPATIONAL DESTINATION BY SKILL LEVEL (PERCENTS)

	OLDS			*FIAT*			*IKA*			*PAL*		
	Un	*Ss*	*Sk*	*Un*	*Ss*	*Sk*	*Un*	*Ss*	*Sk*	*Un*	*Ss*	*Sk*
Community of Socialization												
Rural	75	63	52	68	63	48	24	23	19	69	49	55
Urban	3	6	4	7	6	4	19	31	31	—	—	—
Metropolitan	22	31	44	25	31	48	57	47	50	31	51	65
Totals	100	100	100	100	100	100	100	100	100	100	100	100
	$p = <.05$	$\bar{C} = .2439$		$p = <.10$	$\bar{C} = .2317$		$p = <.20$		$\bar{C} = .1924$	$p = .001;$	$\bar{C} = .3523$	
Father's Occupation												
Farmer	43	42	26	48	33	13	29	20	10	45	26	32
Factory and manual	32	31	38	20	24	32	37	40	52	19	20	19
White collar	12	8	8	17	23	24	25	35	32	31	42	41
Skilled-trade	13	19	28	15	21	31	9	5	6	5	12	8
Totals	100	100	100	100	101	100	100	100	100	100	100	100
	$p = <.20$	$\bar{C} = .2315$			$p = <.01; \bar{C} = .3218$		$p = <.05$		$\bar{C} = .2633$	$p = <.20$	$\bar{C} = .2442$	
Education												
Low	40	42	32	72	59	36	60	30	19	40	8	12
Middle	—	—	—	20	27	33	30	41	34	34	43	34
High	60	58	68	8	14	31	10	29	47	26	49	54
Totals	100	100	100	100	100	100	100	100	100	100	100	100
		N.S.			$p = <.001; \bar{C} = .3563$			$p = <.001; \bar{C} = .5013$			$p = <.001; \bar{C} = .4823$	

manual workers in their social origins, values, politics, and behavior, they resemble manual workers more. While most sociologists de-emphasize the stratification of the working class, I will show evidence of it in all countries and its manifestation in the demographic characteristics of workers, their socioeconomic backgrounds, their labor market experiences, and their current advantages. Whether the stratification increases with industrialization is a debatable question which is examined in Chapter 11. To simplify data presentation of this chapter, the skilled are compared only to the unskilled, and references to individual nations are made only when they deviate from the general pattern.

In advanced industrial societies, skilled workers, compared to the less skilled, tended to be older, and, if married, to have fewer children (U.S. Bureau of the Census, 1960). This pattern held in all countries, except for number of children for IKA workers[4] (see Table 3.8). Since the occupational structures of the four nations have been changing, the amount of direct occupational inheritance is not large. Yet the data reveal that the occupations of workers' fathers influenced the careers of their sons. Compared to the unskilled, the skilled had fewer grandfathers who were farmers and more who were craftsmen or artisans. This pattern persisted everywhere except in IKA. Moreover, except for OLDS, more of the skilled had fathers who were either factory workers, white-collar employees, or self-employed craftsmen.[5] In short, more of the skilled had fathers in nonagricultural occupations who were born locally (especially for FIAT and IKA) or in other urban areas. The greater urban socialization of the skilled was reflected in the larger percentage who resided in the central city rather than on the periphery. Finally, except for FIAT, the skilled were more residentially mobile than other workers.

In all countries, the skilled had large and consistent educational advantages over the unskilled; they received more formal education (see Table 3.8) and, except for India, significantly more vocational or apprentice training. In both India and the United States the skilled were given more on-the-job training.

[4] This pattern held for PAL when the company's but not the researcher's skill classification was used.

[5] These trends were weaker for PAL, but they persisted when the combined skilled and semiskilled were compared with the unskilled.

The less developed the country, the more workers with the least skills were given company training.

Labor market experiences were also stratified everywhere. Skilled workers tended to be older than others on their first jobs, a fact probably reflecting their lengthier education and training. In IKA and PAL, more of the skilled had worked longest in the manufacturing sector while the less skilled had worked longest in the business and service sectors. This pattern was most conspicuous in the less developed economies (Ammassari, 1969). In OLDS and FIAT, slightly more of the skilled had worked in agriculture. Only in Italy did the majority in all skill levels work primarily in the agricultural sector prior to their factory jobs. This unusual pattern probably reflects the sudden burst in demand for industrial labor in northern Italy after World War II.

Not surprisingly, more of the skilled had held skilled jobs prior to their employment in the auto industry. Yet the proportion exhibiting this pattern in Italy and Argentina was surprisingly low.[6] This suggests that the companies selected highly educated workers, gave them intensive on-the-job training, and quickly promoted them to skilled work. This is what any company must do when faced with a critical shortage of experienced skilled labor. During World War II OLDS trained many semiskilled workers for skilled jobs, and this fact partly accounts for the small educational difference among the skill levels. While all the enterprises trained many new employees, PAL and OLDS, the companies which hired more workers from agricultural backgrounds, trained more workers at the semiskilled and skilled levels. All the enterprises engaged in this skill-building function because a majority of skilled employees did not have the requisite experience prior to their employment. While all companies hired a core of skilled workers, it was smallest in the least developed countries, and skilled workers there required the longest on-the-job training.

Whatever the specific routes to mobility, the data clearly show that skilled workers experienced most upward mobility, even more than did sons of farmers. In each factory, three-quarters or more of the skilled experienced upward mobility, and in IKA

[6] Half to eight-tenths of the workers were unskilled prior to employment in auto manufacturing: OLDS, 48 percent; FIAT, 86 percent; IKA, 54 percent; and PAL, 46 percent.

Table 3.8 Social Characteristics of Workers by Skill Levels (percents)

Items	*OLDS*			*FIAT*			*IKA*			*PAL*		
	Skill Levels											
	Un	*Ss*	*Sk*	*Un*	*Ss*	*Sk*	*Un*	*Ss*	*Sk*	*Un*	*Ss*	*Sk*
Median age	28	33	39	32	33	36	27	29	32	33	30	35
Married	90	95	99	70	79	76	69	70	83	79	72	92
Number of children ($\bar{X}$)	1.87	2.29	2.20	1.21	1.04	1.00	1.32	1.63	1.93	1.81	1.37	2.03
Grandfather: farmers	71	71	67	76	63	51	64	64	55	72	47	57
Fathers: urban born	17	20	17	29	31	36	45	57	64	N.A.	N.A.	N.A.
Fathers: skilled	14	19	28	15	21	31	9	5	6	5	12	8
Community of socialization: rural	75	63	52	68	63	48	24	23	19	69	49	35
Education: high	60	58	68	7	14	31	10	29	47	26	49	54
Vocational training	3	2	40	28	24	50	22	14	30	6	6	5
Company training	25	52	51	28	45	33	43	49	45	48	75	79
Age first job: young	7	15	9	13	12	9	48	38	43	20	17	8

TABLE 3.8 (CONTINUED)

	OLDS			*FIAT*			*IKA*			*PAL*		
	Skill Levels											
Items	*Un*	*Ss*	*Sk*	*Un*	*Ss*	*Sk*	*Un*	*Ss*	*Sk*	*Un*	*Ss*	*Sk*
Number of occupations: 4 or more	39	54	54	44	36	36	58	49	46	52	56	55
Sector background: manufacturing	86	93	85	25	33	31	57	71	79	39	65	74
Community residences: 3 or more	44	53	63	50	39	11	42	52	50	66	63	69
Residence: central city	40	40	63	68	72	78	80	78	90	97	84	88
Highest previous occupation: skilled	18	12	41	6	12	10	9	6	26	14	29	41
Unemployment: none	85	75	84	80	85	95	66	50	74	22	24	24
Seniority: 10 years or more	39	85	87	21	34	61	38	47	69	34	27	52
Occupational mobility: up	31	55	78	38	42	80	32	50	64	33	64	79
Occupation liked most: present	50	58	77	31	38	40	60	58	64	48	54	79

two-thirds did so. Contrary to popular belief, the more industrial the country, the less workers suffered from unemployment. Another reward of skill is steady employment. Seniority protects workers against unemployment, and in all countries the skilled had most seniority. Although differences in extent of unemployment were small among workers with different skills for OLDS and PAL, skilled workers everywhere suffered the least unemployment in each age cohort. Finally, skilled workers everywhere earned on the average 50 percent more than the unskilled did. In addition to their advantages in income, prestige, job independence, and upward mobility, the skilled workers everywhere found their work more intrinsically satisfying.

When all the data are evaluated in terms of the number of advantages which the skilled have over the less skilled, it is apparent that in the United States the skilled were most favored; little differences were found for the three other countries. The data point to an apparent anomaly: while less developed countries have more rigid stratification systems, the United States gives most advantages to the most upwardly mobile.

A Note on Caste

Although castes have been officially abolished in India, their influence is still felt in many areas of life. The important research issue here is whether caste affects occupational allocation. The traditional view holds that Muslims are more attracted to industry than are Hindus, who are less materialistic in their life orientations. Among the Hindus, those from high castes are the least attracted to the factory because they consider industrial work to be dirty or polluting (Sovani, 1966:78). In contrast to the traditional view, Lawler (1963:33-51), Niehoff (1959), and Srinivas (1969:65) feel that all castes accept jobs in the industrial sector because it offers higher wages, a view more compatible with the industrialism hypothesis. While data from PAL are limited and not conclusive, they may forecast the future.

The caste system is allegedly unique to Hinduism, but some scholars think that its influence pervades all Indian religions. Many Muslims, Christians, and other non-Hindus are converts from Hinduism or descendants of converts who have not completely shed their caste identities. In view of this, all workers were asked for their caste identification prior to inquiries about their religion, but very few non-Hindus, even when pressed, admitted belonging to a caste. In studying how caste position may affect occupational placement, one must decide whether non-Hindus will be compared to members of individual castes. I decided to examine how religion is related to skill level before examining the effect of caste. (See Table A.2, p. 276.)

When Hindus were compared to non-Hindus, the former were overrepresented among skilled workers, but differences disappeared when the company's skill classification was used. Since the company classification better reflected actual wages, one can safely conclude that the company ignored the religion of applicants in its hiring practices. Yet, compared to the non-Hindus, more of the Hindus were born locally, were younger, better educated, and in jobs requiring above average skill. In time, Hindus with high seniority may earn higher wages than will non-Hindu workers.

When studying caste, one must decide whether to recognize its regional basis, because caste rank may differ in different regions (Srinivas, 1969). Two obstacles made a regional caste analysis difficult. First, Bombay is a large city where migrants constitute a majority of the population, and, second, the PAL sample did not contain sufficient non-Marathas for a separate regional analysis. Since over three-fifths of the Hindus in PAL were born in Maharashtra and the remaining came from scattered regions, I decided to ignore the regional origin of workers.

The distribution of the castes among the skill levels varied according to skill classification used (see Table 3.9). With the re-

TABLE 3.9 THE SKILL COMPOSITION OF THE MAJOR CASTES ACCORDING TO THE CLASSIFICATIONS USED BY THE RESEARCHER AND THE COMPANY (PERCENTS)

	Unskilled		*Semiskilled*		*Skilled*		
Castes	*Res.*	*Co.*	*Res.*	*Co.*	*Res.*	*Co.*	*Totals*
Brahman	28	6	36	62	36	32	100
Kshatriya	27	15	41	41	32	44	100
Artisan	8	—	61	54	31	46	100
Village servants	44	22	26	17	30	61	100
Backward castes	69	6	12	31	19	63	100
Totals	32	11	37	44	31	45	100
N	(55)	(20)	(63)	(75)	(54)	(77)	(172)

searcher's scheme, the two highest castes, the Brahman (priests) and the Kshatriya (warriors) were distributed more or less equally among the skills, the middle Artisan caste was more concentrated among the semiskilled, and the two lowest castes (Village Servants and Backward) were concentrated among the unskilled. However, according to company records, the Brahman were mostly semiskilled workers, the Kshatriya and the Artisans were concentrated most among the semiskilled and the skilled, and the two lowest castes concentrated most among the skilled. In short, the lower the caste, the higher their concentration among the skilled. But by the researcher's scheme, the two highest castes were randomly distributed and the two lowest were concentrated heavily among the unskilled. The picture is clarified by reference to the four major work functions. The Brahman, Kshatriya, and Village Servants were proportionately distributed among the assemblers, machine operators, TIR (test, inspection, and repair), and the skilled craftsmen. Artisans were concentrated in TIR and the crafts, and the Backward castes among the assemblers. Translating these functions into their appropriate skills confirmed the researcher's view of the situation: the company's upgrading of the skills of the lower castes was fictitious. Workers were paid on the basis of seniority, and their jobs were reclassified to justify the increases. The relevance of this for worker dissatisfactions will be examined later.

The bearing of caste characteristics on occupational destination bears further examination. Although caste and rural-urban origins were weakly associated, Backward castes were predominantly rural and Village Servants were predominantly urban. In terms of occupational origin, more of the Brahman had fathers who were white-collar workers, more of the Kshatriya had farmer fathers, while more of the Artisans had fathers who were skilled workers. Yet, fathers of the Village Servants and the Backward castes were well represented in all the occupational categories. Despite the clear association between education and caste status, no caste career mobility pattern appeared, except that more Artisans were downwardly mobile. The Brahman, Artisans, and Backward castes were, on balance, downwardly mobile generationally, while the Kshatriya and Village Servants were upwardly mobile. The upward movement of the Kshatriya probably reflects their rural background and high education, while the mobility of the Village Servants reflects the advantages of their urban background. Downward mobility for the Brahman was inevitable given their white-collar origins. In the factory, the two lowest castes had the decisive advantage of seniority.

A comparison of Hindus from Bombay and Maharashtra with those born outside the state showed that the locally born had acquired more education and were more highly skilled. Local origin was also advantageous to others: a larger proportion of the non-Hindu than Hindus were born in Maharashtra, and their educational level and occupational origins were second only to the Brahman. Since elementary education is not uniformly distributed in India, the educational advantage of the Brahman and Kshatriya reflected their high status and not their youth. However, superior education did not bring them automatic advantages; they were not more upwardly mobile than others. Upon entering the labor market, they did not acquire skilled jobs more readily than others, and, since they lacked vocational training, they had no decisive advantage in factory work. The only clue to management favoritism is the absence of the Brahman and Kshatriya on the assembly line. Management hired most of them to be machine operators or as testers, inspectors, and repairmen. But overall, the older long-tenured lower castes and non-Hindus had the advantages of skill and higher pay.

When PAL began operations as an assembly plant, most of its employees were non-Hindus and lower caste Hindus. Perhaps high-caste workers were not attracted to factory work at the time (Sovani, 1966:78). Later, when production functions were added to the factory, young higher-caste Hindus were employed. Over time, the advantages of tenure were greater than high caste and superior education. However, by over-rewarding longevity and by responding to union pressure to promote nonqualified workers, management created a particularistic wage system that created morale problems. However, in the long run, educated, high-caste workers with long tenure will receive the highest wages.[7]

Discussion

Support for a structural explanation of occupational mobility was provided by data in all four nations. In no instance did em-

[7] Lambert (1963:140) points out a similar finding in Poona: ". . . the occupational and skill structures are obscured by dearness allowances [cost of living], seniority increments and idiosyncratic market considerations, so a skill classification based upon a system of blocks of starting wages was developed."

ployees represent a cross-section of the local labor market; their recruitment was selective everywhere. Although autoworkers occupied a different place in the stratification system of each nation, the most upwardly mobile were rural-reared semiskilled workers whose education was below average. However, from the point of view of occupational destination, the best jobs were held by those who had the greatest opportunities. More skilled workers grew up in urban areas, had fathers who were skilled workers, and were given more formal education and occupational training. The skilled maintained their advantages at every stage of their work life. In conclusion, while similar stratification forces operated in all four nations, differential opportunities tended to exert more influence on work careers in the less industrial countries.

4

Community Origins, Industrial Discipline, and Urban Adaptation

Some Persistent Questions

Both scholars and laymen have worried about the problems rural migrants face when they move to the city. Although the extensive research on this subject has not been systematically organized (cf. Beijer, 1963), two opposing interpretations dominate the literature. The view with widest currency is that migrants are forced to leave their farms and villages because of economic privation. They find city and factory life depressingly impersonal, become disoriented, avoid fellowship with their workmates, and stay away from their jobs[1] (Mayo, 1945). Moreover, migrants fail to become involved in the union, community (Freedman and Freedman, 1956), and politics (Wright and Hyman, 1958). Their conservatism and isolation from working-class political movements (Whyte *et al.*, 1955:39-47), thus reduce labor solidarity. Hoselitz (1955) and others suggest that migrants sometimes transport their rural neighborhoods en masse to the city and that this eases their adjustments. But since they never quite accommodate, they make frequent visits to their villages and return there to retire and die.

Other researchers believe that this dismal picture reflects the pro-rural bias of investigators rather than the reality. The evidence, they feel, clearly shows that, given the opportunity for economic improvement, ruralites happily abandon their villages for factory work. The world-wide tide to the cities constitutes irrefutable evidence that rural loyalties melt before the urban lure of rising consumption (Touraine and Ragazzi, 1961). Migrants adapt easily to factory jobs because work satisfaction is not a salient rural value (Lambert, 1963; Kahl, 1968). Ex-farmers

[1] The literature is contradictory. Whyte *et al.* (1955:42) found that rural workers in industry avoid contact with urban fellow workers and that ruralites have low absenteeism and are rate busters. Feldman and Moore (1960:41-54) suggest that rural workers in less developed societies have low commitment to factory life and high absenteeism.

quickly become tractable employees and loyal members of labor unions. Like urbanites, when the union is strong and participation is encouraged, ex-migrants back the union and attend meetings; where the union is weak, migrants are apathetic. In short, migrants to the city quickly conform to the life style of working-class urbanites (Touraine *et al.*, 1961). In this life style the job is not a central life interest; workers are involved in their families, consumption, and leisure-time activities (Dubin, 1956; Goldthorpe *et al.*, 1968). But they do derive satisfaction from social contacts wherever they occur: in the factory, union, neighborhood, or community (Wilensky, 1961a). Though workers fail to become organizationally active in the community, they adapt to the urban-industrial way of life with little strain (Wyatt and Marriott, 1956).

This second and more optimistic view of urban and industrial adaptation, though based primarily on research in industrialized countries conforms to results found in Puerto Rico (Reynolds and Gregory, 1965) and other less developed areas (Inkeles, 1969a). The industrialism hypothesis holds that especially in impoverished countries, rural migrants to the city feel so fortunate to have secure factory jobs that they become overcommitted to them (Feldman and Moore, 1960; Lambert, 1963). The development hypothesis takes a more moderate view; rural-bred workers, socialized to follow tradition, need time to adjust to the strains of factory and city life, but they do so within a few years. The lower the level of the nation's industrialization, the more traditional is the rural milieu, and the more difficult and delayed is the migrant's adaptation to urban living and industrial discipline (Smelser 1963a:43-48).

Conflicting research findings may result from the failure of scholars to control for technological variables. Not infrequently workers of rural and urban origin have been compared even when they worked in different industries. The present research reduces the confusion by studying workers only in automobile manufacturing. Only small national and rural-urban differences in accommodation are expected because I control for technology and urbanism, but rural-urban differences, should they occur, will be larger in the less developed societies.

Wherever automobile factories are found, a relatively complex urban-industrial environment also exists. In order to make automobiles, factories must impose the same discipline on all employees regardless of their community origins (Inkeles, 1969a; Karsh and Cole, 1968). Exposure to the same bureaucratic organization, exposure to the same technology, and exposure to similar work arrangements force workers to take on a similar interpretative scheme of events. Insofar as life in the family, union, and community are affected by what goes on in the factory, workers from both urban or rural background should be affected similarly. Undoubtedly urban-bred workers have been exposed longer to an environment in which the factory, the union, and its problems are part of everyday existence. But these problems are not so complex or so unique that workers reared in rural areas cannot quickly grasp their essence. They have already made the biggest transition by moving to the city. Since they have no independent frame of reference to respond to the factory problems and problems related to it, they will be disposed to adopt the views of urban workers.

Rural-urban Backgrounds

Four variables reflecting rural or urban backgrounds were available: community of birth, community of socialization, father's occupation, and main sector (agricultural, manufacturing, or service) of work experience. All these indicators have shortcomings. With community of birth, the problem is to decide what population size makes a community rural-like or urban. Then, larger communities in the hinterland may be more rural than small towns near large cities. Moreover, a person born in a small community may leave it early in life. Community of socialization (where the respondent lived between the ages of ten to fifteen years) is superior on this score, but a decision still must be made on community size and location. Ascertaining the size of community when the respondent was an adolescent is tedious even when good census data are available—which was not the case for Argentina and India. Not only did we have to guess community size, but we had to take cognizance of the fact that size has different organizational consequences in different countries. In the United States and Argentina, a community of ten thousand is not considered to be rural, while in Italy and to some extent in India, farmers live in large communities which include urban services and factories. Father's occupation is a serviceable indicator, but it is crude; all nonfarm occupations are assumed to be urban. The ideal indicator of exposure to rural or urban values is adult work experience in the agricultural or industrial-service sectors. In the United States and Argentina too few workers had been employed in the rural sector to make the indicator useful.

The best indicator is probably that which is most highly related to all the others. On this basis father's occupation was selected because it was related to all indicators as measured by the chi-square test (below .05). Moreover, the correlations (corrected coefficients of contingency) were moderately high, ranging from .213 to .561. Community of socialization and sector of work experience were not related in the United States and in India, and neither were community of birth and sector experience. So, whatever its shortcomings, father's occupation was selected as the best indicator; sons of farmers were considered as rural, and other sons as urban. However, all the other indicators were also examined for their association with the accommodation variables. Whenever father's occupation shows no association with the variables and other indicators do, this will be reported, thus biasing the analysis in favor of the hypothesis that community background affects adaptation.

Community Origins and Paths to the Factory

Despite national differences in urbanization (United States is high, followed by Italy, Argentina, and India), sons of farmers constituted the same percentage (one-third) of the labor forces in three factories, except in IKA where they were one-fifth. Therefore father's occupation cannot be used to rank the rurality of the four labor forces. Exploring other variables, we note that all farmers' sons had grandfathers who were farmers, in contrast to half of the non-rural sons.[2] Expectably, the majority (50 to 90 percent) of rural-origin workers were reared in rural or small communities, but many of the non-rural (17-48 percent) were also reared in such communities.[3] However, although nearly half of all employees grew up in rural areas, fewer than one-fifth had worked primarily in agriculture prior to locating factory jobs.

[2] The Indian case is a deviation and the statistic is less reliable because one-fifth of the respondents did not know their grandfather's occupation.

[3] For India, "the village" was accepted as defining a small community; for Argentina, a town under 15,000; for Italy, under 20,000; and for the United States, under 5,000.

Importantly, half the rural Italians and three-tenths of the rural Indians had been agricultural workers, in contrast to about only one-tenth of the Americans and Argentinians (see Table 4.1). Since rural socialization and agricultural employment were more closely associated in Italy and India than in the United States and Argentina, the labor forces of the factories should be ranked from high to low in rurality as follows: FIAT, PAL, OLDS, and IKA. Therefore, the impact of community background on accommodation should have greatest force in the first two countries.

In all samples, the great majority of non-farm sons became factory employees immediately upon completing school. Only in Italy did more sons of farmers work in agriculture rather than in manufacturing prior to their employment in FIAT. The move from rural background into the factory was greatest for the OLDS sample, where 88 percent of the sons of farmers entered the factory on their first jobs. Only for the IKA sample did as many rural migrants work in the service sector as in the manufacturing prior to becoming IKA employees. These results do not confirm reports that rural migrants in developing societies typically move into the service sector before locating factory jobs (Moore, 1966:203). However, the data do show that the less industrial the country, the more urban-born workers find jobs in the service sector prior to locating industrial employment (cf. Ammassari, 1969).

Community origin affected both the present residence and occupational mobility of workers. The more industrial the country, the more did workers with farm backgrounds reside outside the central city. This trend probably reflects the greater availability of transportation in the more developed societies. Data in Table 4.2 clearly reveal that farmers' sons experienced more generational upward mobility than urban workers' sons. Although this finding is largely a result of classifying farmers as unskilled, this classification did not affect career-mobility patterns the same way because they were smaller than generational mobility differences in all countries. Although no statistically significant differences in the skill level of occupational destinations appeared by rural-urban origins for OLDS and PAL workers, in all countries more of the urban born were skilled. Why is community origin not related to skill level the same way in all countries?

Job tenure, age, and education are related to skill level in complex ways. In all four factories sons of farmers were older, had

Table 4.1 Occupational Derivation, Community Backgrounds, and Sector Experiences, According to Fathers' Occupations (in percents)

	United States OLDS		*Italy* FIAT		*Argentina* IKA		*India* PAL	
	Fathers' Occupations							
Variables	*Farmer*	*Other*	*Farmer*	*Other*	*Farmer*	*Other*	*Farmer*	*Other*
Grandfathers' occupation: farmer	87	54	95	48	87	56	96	36
Fathers' occupation: farmer (total)	100[a]	37	100[a]	34	100[a]	21	100	34
Community of birth: state or province	73	48	53	32	54	19	48	37
Community of socialization: rural or small town	91	45	89	48	46	17	78	37
Community of residence: outside central city	72	51	40	22	25	17	8[a]	11
Number of communities resided: two or more	52	52	52	30	68	41	73[a]	62
Main sector of previous work experience:								
Agriculture	12	2	47	5	8	—	31	5
Manufacturing	88	90	44[a]	81	44	72	44	65
Business and service	—	8	10[a]	14	47	27	25	30
Number of cases	86	153	101[a]	195	57	213	89	168

[a] Probabilities of the χ^2 test not below the .05 level. The tests of significance are based on the original tables which had four categories: sons of fathers in farming, manufacturing, white collar, and skilled trades.

TABLE 4.2 DEMOGRAPHIC AND OCCUPATIONAL CHARACTERISTICS ACCORDING TO FATHERS' OCCUPATIONS

	United States OLDS		*Italy* FIAT		*Argentina* IKA		*India* PAL	
	Fathers' Occupations							
Variables	*Farmer*	*Other*	*Farmer*	*Other*	*Farmer*	*Other*	*Farmer*	*Other*
Age: mean	44.6	39.7	36.5	35.2	33.4[a]	29.5	35.2	32.0
Married: percent	95	93	80	73	79	69	90[a]	77
Number of children: mean	2.7	2.7	1.0	.7	1.7	1.5	2.4	2.1
Education: mean years	8.9[a]	10.2	3.8[a]	5.7	4.7[a]	7.1	5.5[a]	6.9
Age at first job: mean	17.3	17.4	13.4	14.0	15.5	15.2	17.6	17.5
Unemployed: mean months	5.6[a]	3.8	2.4	3.8	2.7	3.7	4.8	5.2
Jobs: mean	3.9	3.8	3.6	3.3	3.9	3.6	4.0	3.9
Job tenure: mean years	14.0	12.4	7.6[a]	9.7	5.8	4.9	8.3	8.3
Occupational skill: percent								
Unskilled	31	26	44[a]	24	63[a]	44	48	31
Semi-skilled	56	52	49	52	32	37	29	43
Skilled	13	22	7	24	5	19	23	26
Upward career mobility: percent	58[a]	47	59[a]	43	53[a]	40	65	51
Intergenerational mobility: percent above father	69[a]	26	58[a]	26	51[a]	21	79[a]	16
Number of cases	86	153	101	195	57	213	89	168

[a] Probabilities of differences statistically significant by χ^2 test at or below .05 level. The tests were based on the original tables and not the classes reported in this table.

longer job tenure (except in Italy), but less education than did sons of urban workers. They did not differ for marital status, number of children, age at first job, number of jobs held, and duration of unemployment. Special conditions in each plant account for the composition of skilled workers. In the United States, some older rural-born workers with long tenure and low education were trained for and promoted to skilled jobs during and after World War II when skilled labor was scarce. In India, the oldest workers were the original employees of the plant when it only assembled automobiles, but as other operations were added, many long-tenure workers were promoted to skilled jobs. Newer and often better-educated workers were hired at the semiskilled level. In Italy and Argentina, where production was expanded very rapidly, the companies hired mostly young workers. Better-educated urban applicants filled the better jobs while less educated rural applicants went into unskilled and semiskilled jobs. In short, rural migrant workers were older, less educated, less skilled, but more upwardly mobile than urban workers. The employment histories and occupational destinies of the two sets of workers differed in each plant in response to varying economic conditions and personnel policies.

Adaptation to Work Organizations

A central objective of this research was to ascertain whether industrialization affects worker adaptation in a widening set of social systems: family, work-group, labor union, neighborhood, community, and the nation. In this chapter the question is whether community origin affects system participation differently in each country. To support the industrialism hypothesis, community origin should have little effect on system involvement anywhere. To support the development hypothesis, Indian rural migrants should differ most from the urban-born, and rural-urban differences should decrease in Argentina, Italy, and the United States. In addition, the Indians should be the least involved in systems associated with urban-industrial life (work-group, union, city, and nation), and the Americans the most involved.

The first set of adaptations to be examined center on work: factory employment, occupation, job routines, work-group, and the labor union. A series of questions focused on sector preference (agricultural, manufacturing, or office). Respondents were asked, "With the same hourly pay would you prefer to work on a farm machine, office machine, or a machine in a factory?" Data in Section 1 of Table 4.3 reveal that three-fifths of the Indian and United States workers preferred the agricultural sector, while the Italians and especially the Argentinians preferred the factory. The office sector was preferred least in all countries. A curvilinear trend was apparent: workers in the least and the most industrialized nations (the Indians and Americans) most preferred the agricultural sector; those in the rapidly industrializing nations, Italy and Argentina preferred the industrial sector (Gale, 1968). Rural-urban differences were strong for the United States and India (corroborated by data on situs work experience), moderate for Italy, and absent in Argentina.

To supplement data on sector preference, workers were asked to indicate which sector was the most desirable, respected, necessary, satisfying, and monotonous to work in. An index of industrial sector preference was constructed by adding the number of times skilled factory work was selected first on each of the five dimensions (see Table 4.3). The same curvilinear relationship appeared, i.e., the United States and Indian workers preferred the industrial sector least while the Italians and Argentinians preferred it most. Although the urban-reared preferred the industrial sector more than the rural-reared in all nations, in no instance was the difference statistically significant.[4]

Sector preference indexes were also constructed for agriculture and the office. The same curvilinear relationship of strong preference for the agricultural sector appeared for the American and Indian workers. Rural migrants in the United States most strongly preferred the agricultural sector, but rural-urban differences were smaller for Italy and India and nonexistent for Argentina. The office sector, except for Argentina, was preferred slightly more by urban respondents in all nations (see Table 4.3). The conclusion is clear: although more of the rural Italians and Indians than others had worked in agriculture, more of the Italians than Indians rejected the agricultural sector and preferred factory work. Despite obvious differences in background, the responses of Americans and Indians were more alike than were those of workers of other nations.

[4] It was largest for India, and the rural-urban difference was statistically significant for sector work experience.

TABLE 4.3 SECTOR AND OCCUPATIONAL ADAPTATION ACCORDING TO OCCUPATIONS OF FATHERS (IN PERCENTS)

	United States		*Italy*		*Argentina*		*India*	
	OLDS		*FIAT*		*IKA*		*PAL*	
	Fathers' Occupations							
Variables	*Farmer*	*Other*	*Farmer*	*Other*	*Farmer*	*Other*	*Farmer*	*Other*
1. Sector preferred								
Farm machines	58[a]	32	34	16	4	8	62	48
Factory machines	36	56	52	54	77	69	30	42
Office machines	6	11	14	30	19	23	8	10
Total	100	99	100	100	100	100	100	100
Sector preference indexes (medians)								
Agricultural (0–5)	3.12[a]	2.25	1.68[a]	1.09	1.71	1.74	3.61	3.00
Industrial (0–5)	1.65	1.78	2.38	2.67	3.29	3.57	1.89	2.37
Office (0–5)	1.49	1.71	1.88	2.14	1.67	1.63	1.52	1.70
2. Occupational adaptation items[b]								
Occupational evaluation: good	62	70	50	52	89	89	54	61
Occupational satisfaction: high	77	78	73	72	91	87	62[a]	65
Job routines: satisfied	81	83	76	77	83[a]	87	65	70
Reference group occup. eval. index (0–5); high (5)	60	62	87	80	91	85	84	86
No thought of having different job	42	30	50	44	56	53	N.A.	N.A.
Not planning a job change	87	83	87	81	77	69	N.A.	N.A.
Index of overall work satisfaction (0–6); high (5,6)	34	36	38	35	42	47	21	10

[a] Probability of the χ^2 test at or below the 5 percent level. The χ^2 tests for the sector preference indexes were computed for the original tables.

[b] Items were selected from complete tables to illustrate trends, but sometimes the trends are not discernible from the item

Whatever their sector preferences, all respondents were working in factories and had to respond to the physical conditions of work and their job demands. They were asked to evaluate these conditions and to provide information on occupational aspirations for themselves and their children. They were also asked how friends, family, and fellow workers evaluated their occupations.

The findings were consistent for all questions: very few differences appeared between the sons of farmers and the sons of other workers.[5] Data in Table 4.3 show that in all nations workers adapted to their occupations well: almost all were satisfied with the physical work environment; half or more evaluated their occupations as good; two-thirds or more indicated high occupational satisfactions; and an equal ratio were highly satisfied with their specific job routines. The Argentinians appeared to be best adapted, followed by the Americans, Italians, and the Indians. In all of these measures, the sons of farmers and the others responded almost identically. However, except for Argentina, when they responded differently, those with rural backgrounds seemed less adapted to their occupations than the urban-reared (see Table 4.3).

Three-fifths or more of the respondents felt that their families, friends, and co-workers evaluated their occupations positively. The Americans had the lowest reference group support, probably because factory jobs have less prestige in a service than in an agricultural or manufacturing economy. Moderately high work adaptation was also reflected by the high proportion (more than 70 percent) of workers who were not planning to change jobs. Rural-urban differences were slight for all indicators of occupational accommodation. However, the Italians and Argentinians with rural backgrounds had slightly higher reference group support for their jobs than did the urbanites and in all countries, a smaller ratio of the rural than urban-born had considered or planned to change jobs.

In conclusion, rural-urban differences in occupational adaptation were typically small. Since workers from rural backgrounds were more concentrated in lower-skill jobs than urban workers, we might expect them to be less happy about their jobs. This was

[5] Of the 96 chi-square tests of significance, nine had probabilities at or beyond the .05 level. This is twice the number expected by chance, but the distribution of these "deviations" followed no discernible pattern.

not the case, probably because those with rural backgrounds had other sources of satisfaction: they had experienced considerable upward occupational mobility, they had steady well-paying jobs which they did not want to lose (cf. Browning, 1971), their reference groups evaluated their jobs highly, and they had relatively low occupational aspirations.[6]

Interaction at Work and in the Union

Some observers (Wyatt and Marriott, 1956) assert that factory workers find little satisfaction in their jobs but make up for its lack by developing congenial informal work-groups. Rural migrants are allegedly less integrated into these informal work-groups (Whyte *et al.*, 1955; Blood and Hulin, 1967), but the data from this study show that workers from both rural and urban origins are highly and equally satisfied with their work mate contacts (see Table 4.4). I devised measures of the amount and quality of social interaction which workers had inside and outside the factory. No consistent differences were found between the two groups in the number of work mates they talked to while working[7] (see Table 4.4). Since this finding could reflect the uniform constraints on interaction imposed by the technology and ecology of work organization, I attempted to measure the extent that workers took advantage of the potentialities for interaction: the number of people with whom they *actually* conversed divided by the number in their work space with whom they could talk. Although considerable variation appeared in the four plants on the amount of potential interaction workers realized, differences between the rural- and urban-reared workers were rare and small.[8] Finally, two other indexes were constructed, one which summarized the totality of interaction with fellow workers inside and outside the plant, and the other, the quality of that interaction. Again, no significant differences were found between the rural- and urban-origin groups within the factories, and no

[6] Over half of the sons of farmers expected their children to become white-collar workers. Although the percentage was higher for other workers, the difference was not statistically significant.

[7] The trend in OLDS and FIAT was for sons of farmers to talk to fewer work mates than sons of nonfarmers, but those who had agricultural work-experience talked to more of their mates.

[8] Weak tendencies appeared in the PAL data for farmers' sons and those with work experience in agriculture to interact less with their work mates.

TABLE 4.4 INTERACTION AND PARTICIPATION IN VARIOUS SOCIAL SYSTEMS ACCORDING TO FATHERS' OCCUPATIONS (IN PERCENTS)

	United States OLDS		Italy FIAT		Argentina IKA		India PAL	
	Fathers' Occupations							
Variables	*Farmer*	*Other*	*Farmer*	*Other*	*Farmer*	*Other*	*Farmer*	*Other*
Work mate contacts: satisfied	86	90	91	84	91	94	75	74
High workplant interaction: talks to 11 or more workers	34	36	28	40	14	9	25	38
Index of quality of work mate interaction (0–6); high (3–6)	57	59	49	47	47	54	66	67
Potential of work-place interaction realized (0–1.0); high (1.0)	60	53	38	32	62	76	37	40
Union participation (0–2); high (2)	31	36	20	24	65	58	44	33
Union evaluation (0–3); high (3)	49	59	26[a]	39	32	38	63	61
Conservative and radical index (0–3); liberal and radical (2,3)[b]	24	27	61	70	25	32	39[a]	48
Family interaction (0–3); high (2,3)	41	42	61	69	51	42	26[a]	40
Neighborhood involvement (0–6); high (5,6)	42	40	28	25	25	30	28	35
Community involvement (0–3); high (3)	23	29	16	24	33	25	40	45
National involvement (0–3); high (3)	27	36	10[a]	23	28	36	45	54
Index of total social interaction (0–8); high (5,6)	28	31	17	22	11	19	28	43
Anomie (normlessness) index (0–5); means	2.6	2.5	3.2	3.2	1.8	1.7	3.1	3.4

[a] Probability of the χ^2 test at or below the .05 level.
[b] For the United States, liberals only.

pattern was evident among the countries (see Table 4.4). However, whenever tendencies appeared in the data, ruralites, contrary to common belief, had slightly more work mate interaction.

Rural migrants in industry allegedly distrust labor unions because they feel that unions are too ready to resort to violence and conflict (Whyte, 1944; Whyte *et al.*, 1955). In addition, unions are too inclined to support liberal or radical political parties and force their members to endorse these parties and their ideologies. While some studies have found that uprooted rural migrants tend to be politically radical (Leggett, 1963a:690), most researchers have concluded that ruralites in industry tend to be politically apathetic or conservative (Lipset, 1960:101-114). I compared the rural and urban groups on their attitudes toward unions, their participation in union affairs, and their political beliefs. Although sons of farmers were slightly less enthusiastic about labor unions than sons of nonfarmers, the great majority of both groups in all countries accepted and supported the unions. No statistically significant differences appeared between the two groups in their interest in unions, in union participation, or in their judgments of unions acting as collective bargaining or as political agents (see Table 4.4). The fact that employees of rural origin participated slightly less in union affairs reflected their lower occupational status rather than antipathy toward unions.

Finally, I constructed indexes of political ideology (conservatism-radicalism), which were adapted to the political circumstances of each nation. Important differences in the scores of the rural and urban groups appeared only in India. While workers from rural backgrounds were slightly more conservative than others, sons of farmers in India were politically more polarized than sons of urban workers, but the rural, on balance, were more conservative.

Social Interaction Outside the Factory

To complete the profiles of social interaction, I constructed indexes of worker involvement in family, neighborhood, community, and broader societal systems. Traditional theory (e.g., Toennies, 1957:33-65) suggests that people with rural backgrounds should be more involved in family and local neighborhood activities, while urbanites should be more involved in city and national affairs. An index of family involvement was constructed by ascertaining the extent that a number of activities (visiting, vacationing, going to a bar, discussing politics) were performed with family members or relatives. Family involvement was highest in Italy and Argentina, but only in India was there a rural-urban difference, but in a direction opposite to that expected: urbanites were more family-involved than were the rural group. This finding probably reflects differences in family arrangements. Since more of the rural group had left their families back in the villages, interaction with them was necessarily reduced.

In a measure of neighborhood involvement (desire to continue living in area, number of friends in neighborhood, number of relatives in neighborhood, attendance at neighborhood meetings, and naming neighborhood problems), the Americans were found to be the most highly involved. Although national rural-urban differences were minuscule, consistent low-order differences appeared for involvement in community and national affairs. Workers from urban backgrounds were more informed about community and national events, joined more organizations, and discussed more local and national issues. However, in only one instance (Italy, for national involvement) were differences between origin groups large enough to be statistically significant (see Table 4.4).

Finally, two summary measures were constructed: an index of interactions in all social systems both inside and outside the factory,[9] and a Guttman scale of normlessness anomie. No important rural-urban differences were found in total interaction scores, but in all four nations rural workers had slightly lower scores (see Table 4.4). Finally, PAL and FIAT workers apparently saw their societies as more anomic than did OLDS and IKA workers, but again no strong differences appeared between the urban and rural origin groups in each society.

Problems of Classification, Tenure, and Caste

This tedious exercise has shown that the rural or urban background of workers had little effect on their participation and accommodation to work and nonwork social systems. Yet upon ex-

[9] Key (1968) has the best analysis of the effect of rural-urban backgrounds on participation in different social systems and communities.

amining the direction of the weak trends when they appeared, I found that workers from rural backgrounds were slightly less involved in community and national affairs than urban workers. However, since sons of farmers were less educated and occupied lower-skilled jobs than sons of nonfarmers, and since workers of lower socioeconomic status are typically uninvolved in nonfamily social systems, rural-urban differences in system involvement undoubtedly reflect the socioeconomic status of the two groups.

Since the above analysis was based on a crude dichotomy (sons of farmers and nonfarmers), I reran the analysis for different systems of classification. First, I compared sons of farmers to each of four occupational groups: sons of industrial, white collar, artisans, and service workers. Second, I tried to take into account the fact that, over time, participation in social systems tends to increase. Finally, since in India the adaptation of the rural workers could be distorted by caste, I examined adaptation by caste.

First, I attempted to ascertain which urban occupational category (factory, white collar, artisan, or service) was most similar or dissimilar to sons of farmers on all the items discussed in this chapter. In every instance but one, sons in at least one of the urban categories responded exactly as did sons of farmers. The one exception was sector preferences, where the sons of farmers preferred agriculture far more than did workers in the other categories. An overview of the findings shows that in Italy sons of farmers responded to most items most like the sons of factory workers, and in Argentina, most like the sons of artisans. But no pattern appeared in the United States and India. When I searched for the greatest differences in the responses of workers in the various occupational categories, I could find no pattern for Italy and India; in the United States, sons of artisans were most unlike sons of farmers; in Argentina the sons of white-collar workers differed the most. In all countries, rural migrants differed most from one other group in work-plant interaction, sector preference, union evaluation, and community involvement. However, since both the direction of the differences and the occupational categories involved varied by society, the results pointed to a random pattern. Overall, no discernible pattern among origin groups or nations resulted from this re-analysis.

Second, sons of farmers and nonfarmers were compared for their patterns of involvement in work and other social systems according to their length of employment in manufacturing. Except for PAL workers, industrial tenure was positively associated with job, work, and reference-group satisfaction, but not to sector preference. Although workers with long tenure interacted somewhat more with their work mates and participated more in the union than did low-tenured workers, no tenure differences appeared for family, neighborhood, community, and national involvement, nor for quality of work mate interaction, political ideology, or anomie. We may conclude that the pattern of system involvements became quickly established in all countries.

The critical question then is whether the community origins of workers, over time, has a lessening effect on their work adaptation and system involvement. Surprisingly, in the less developed nations (Argentina and India) the seniority of rural-born and urban workers was not related to anything: neither to their demographic attributes, career patterns, or system involvements either inside or outside the plant. For Italy and the United States, a few *small* differences appeared. In Italy, highly tenured urban workers tended to be older than sons of farmers, to be less educated, to spend more time in the manufacturing sector, and to be better adapted to their work, job, occupation, and the factory. They were more involved in the work group, family, community, and nation, but less involved in the union. They also had higher anomie scores. In short, except for low union involvement, FIAT's urban employees approximated the classic model of the proletariat. In OLDS, the *low*-tenured urban workers followed the FIAT pattern with few exceptions. They had spent more time in industry than had the rural; were more adapted to their occupations; more involved in the work group, family, neighborhood, and national systems; but were more liberal politically and had lower anomie scores. In sum, tenure did not distinguish between the adaptation of rural and urban workers in the less developed societies but it had small effects on older and younger urban workers in the more industrialized countries.

Finally, caste in India was not strongly associated with the rural-urban background of employees. Although higher-caste workers were younger and better educated than lower-caste, the latter had higher seniority and were more heavily concentrated in the higher skills—according to the company's classification. No statistically significant differences among castes were noted for work, occupation, job, and sector adaptation. Neither were there important or consistent caste differences in social systems involvement inside or outside the factory.

In conclusion, different modes of classifying respondents into urban and rural categories did not affect the general findings. Tenure was not related to work adaptation or to patterns of system involvement for workers from rural or urban backgrounds in the two less industrialized countries; background and tenure effects were inconsistent in the other two countries. In India caste was not associated with community origin or with patterns of work and social systems involvement.

Discussion

According to the industrialism hypothesis, workers with different community backgrounds accommodate quickly to factory and urban environments; the development hypothesis stresses that workers from rural areas experience more difficulty, especially in the less industrialized societies. This study found relatively few community background differences within nations in patterns of adaptation. Moreover, community background did not produce increasing effects in the nations which were most rural, as developmental theorists suggest that it should. While Indian workers seemed less adapted to factory work than employees in the other factories they were as heavily involved in factory and other social systems.

These negative findings do not prove the industrialism hypothesis of direct technological effects. Perhaps rural migrants are more highly educated than rural nonmigrants and more inclined to accept urban-industrial patterns prior to migrating (cf. Conklin, 1973 for India). Recent American studies have found only small differences between workers from rural and urban backgrounds (Blau and Duncan, 1967:277-292). Fuller (1970) suggests that this is so because urban-industrial patterns have permeated the hinterland and wiped away traditional patterns. A similar process may be at work in less industrial countries. But where dual economies exist (Boeke, 1942; Higgins, 1956), urban-industrial influences may not penetrate the hinterland, and rural migrants may not experience much anticipatory urban socialization. In this study the Indians were less adapted to industrial work than other employees, but even in India rural-urban differences in social system involvements were small. Most of the Indian migrants, like those in the other countries, came from nearby regions which were exposed to urban-industrial influences.

Inkeles (1969a) has proposed an explanation which applies to migrants from both nearby and distant areas. In a study of modernization attitudes, he demonstrated that the factory functions as a school for modernization (1969a:212-216) almost as effectively as formal education itself; the longer workers are employed in factories, the more modern their outlook. Thus workers born in distant rural areas would acquire an outlook similar to the urban-born after working in a factory for a number of years. In the present research, rural migrants in Argentina had less factory experience than workers in other countries, but even the Argentinians had an average tenure of five years, which is sufficient to expose them to factory modernizing influences. Inkeles' hypothesis also fits highly industrialized societies. Thus, Zimmer (1955:223) demonstrated in Flint, Michigan, that the social and political participation rates of rural migrants were low during the first five years of urban residence, but increased rapidly after that time. Touraine and Ragazzi (1961) found that French automobile workers from farm backgrounds adapted quickly to industrial situations, although they did not become fully integrated into the industrial way of life. And Signorini (1970) and Barbichon (1967) found that Italian industrial workers from rural areas experienced some adaptation difficulties during the first year, but became well adapted after that.

Results from this research and hints from other studies suggest that rural and urban backgrounds do not affect worker accommodation in the factory or their social system involvements outside of it, especially after employment in manufacturing for five or more years. There is less evidence that community origins have differential effects on working-class social or political solidarity. Technological relations in the factory probably exert more influence on occupational adjustment, on social interaction with work mates, and on system participation. The next chapter examines how factory technology influences worker behavior inside and outside the plant.

5

Technology, Machines, and Worker Behavior

European social theorists have long lamented that industrial technology has produced generations of unhappy and alienated industrial workers. The presumption is that the spread of technology to newly industrializing societies will have similar effects. Thus, workers who have been reared in the secure and intimate milieu of traditional communities will suffer greatly and will need much time to become committed industrial employees (Moore and Feldman, 1960). But recently, some studies have shown that new industrial workers are not traumatized by their contacts with machine technology; most seem to accommodate quickly to factory discipline (Lambert, 1963; Reynolds and Gregory, 1965; Karsh and Cole, 1968). Inkeles (1969b) suggests that the factory may actually facilitate workers' adaptation to urban life by teaching them discipline, providing them with opportunities to meet people outside their traditional circles, and training them to participate in formal associations such as labor unions.

Studies of employees in older industrial societies have indicated that they have succeeded in creating intimate informal organizations in response to impersonal factory bureaucracies (Roethlisberger and Dickson, 1947; Jaques, 1952; Whyte *et al.*, 1955), and that they have formed integrated social circles in the neighborhood and community (Janowitz, 1952; Keller, 1968; Booth and Edwards, 1972) in response to urban anonymity. Apparently, in contrast to newly industrializing societies, wherever industry is well established, workers should have intimate and extensive relations with their work mates both inside and outside the plant. This firmer social network (Craven and Wellman, 1974), should reduce their anomie below that of new industrial employees (Smelser, 1963a:44). Yet until comparative studies measure the extent and quality of worker interaction in and outside of factories in societies at different industrial stages, generalizations about how technology affects workers' lives are mere speculations.

This study takes the position that the social effects of technology in different countries are similar whenever their technologies are similar. Employees who are exposed to similar technologies, whatever their previous socialization, local culture, and level of societal industrialization, will respond similarly to that technology.[1] The spatial distribution of machines, the skills required to operate them, and the organization necessary to coordinate workers strongly influence social interaction (Sayles, 1963; Blauner, 1964; Meissner, 1969). These factory-based patterns of interaction may also influence the community behavior of employees. Thus, the amount and type of work interaction, rather than the recency of exposure to industrial organization, shapes how workers respond to the factory and the city. The ideal test of these ideas requires that workers be studied in factories which have identical technologies but are located in societies with different cultures and at different industrial stages. This ideal is impossible to achieve but it is roughly approximated in this research.

Technological Characteristics of the Plants

The distribution of skills and work operations of the four plants reflects their different technologies (Table 2.4). OLDS made the most complex product, had the most complex technology, the most automatic equipment, and employed relatively more personnel in quality control (semiskilled workers in test, inspection, and repair). Although FIAT's technology closely resembled OLDS's, FIAT manufactured more of its parts and thus had more machine operators. IKA produced several models of cars and trucks, but was not so automated nor so mechanized as OLDS or FIAT. PAL's technology was the most primitive; more parts were fabricated by skilled rather than semiskilled workers, and more skilled workers were engaged in test, inspection, and repair (TIR) functions.

The experimental and machine maintenance departments of the four factories were most alike in their skill composition, but the physical distribution of their machines differed. OLDS

[1] Other sociologists of a technological persuasion (Woodward, 1958; Burns and Stalker, 1961; Zwerman, 1970) have investigated how technology affects the social organization of industry and management behavior.

had the largest and most dispersed machines. Machines in the parts-making departments were relatively uniform in size and placement in all four factories, but the final-assembly departments differed more than the other departments. OLDS's assembly lines were the longest, most rationalized, most mechanized, and moved the fastest. Except for adjustments necessary to assemble smaller vehicles, FIAT's lines were the same as OLDS's. IKA had several crowded, slow-moving lines for different types of vehicles, and workers were periodically rotated from one line to another. PAL's crowded assembly lines were not mechanized.

Technological Constraints and Interactional Opportunities

Since this research hypothesizes that technology heavily circumscribes worker behavior, the technologies of four major operations are described in detail. Technology includes (1) the tools and machines required for the operations, (2) the routines required to operate the machines, and (3) the design of the work flow. Departments in mass production industries are typically technologically homogeneous, and this homogeneity is particularly likely in automobile manufacturing (Woodward, 1958). Two types of technologies are not widespread in the automobile industry: hand-tool or craft technology and automated continuous-process technology. Although the two technologies differ greatly, both permit considerable interaction and encourage the formation of solidary work groups (Blauner, 1964). In sharp contrast, the technology of textiles, for example, disperses its workers widely, thus restricting their interaction and social solidarity.

Automobile manufacturing lies in the middle range of technological complexity. At the simpler levels (e.g., house construction), since tools are simple and machines have multiple functions, workers can control the details of their work and their interaction (Meissner, 1969:244-245). In the middle range of technological complexity, employees have less freedom of movement because semiautomatic and automatic machines need constant attention. However, since the workers' routines seldom fit machine requirements perfectly, workers and supervisors fight over many things. Workers mainly push for the freedom to control their machines; supervisors push workers to adapt to the machines' rhythm, thus increasing production. At the highest point of technological complexity, where processes are automated (as in the oil refinery), production is not under worker control; workers monitor meters, turn valves, and engage in nonrepetitive tasks. These tasks permit considerable work autonomy and freedom for social interaction (Shepard, 1971:26-28).

Though automobile manufacturing is in the middle range of technological complexity, a considerable range in worker-machine relationships exists among the four main processes (see Table 5.1). Assemblers are confined to a restricted area, and their work requires only superficial attention and almost no communication. However, frequent communication does occur, although the noise makes it brief and intermittent.[2] Operators of semiautomatic machines which make identical pieces live in a technological environment almost as restrictive as that of the assembly line. Since production targets are high and machine-tending is almost continuous, freedom to move about is restricted. This condition and the high noise level greatly restrict communication. TIR operations permit more interaction than assembly and machining. While constraints on some inspection operations are high (e.g., routine quality control along an assembly line), other operations (inspecting carburetors) permit continuous conversation. Workers who test motors and make minor repairs have intermittent opportunities to move about and talk. Finally, workers who make and repair machines have considerable freedom to communicate. They can set their machines or leave their work stations to converse with others for considerable periods. Occasionally their work requires consultation with supervisors and fellow workers, but continuous interaction is rarely required. The autonomy which this type of technology permits encourages the formation of solidary work groups (Sayles, 1963; Lipset *et al.*, 1956). In summary, data in Table 5.1 show that craft work, TIR operations, machining, and assembly constitute an increasing hierarchy of technological constraints on interactional opportunities.[3]

[2] Touraine (1955) warns that it is easy to exaggerate the uniformity of social interaction on assembly lines; great variations are possible within and between lines.

[3] This hierarchy is similar to Tudor's (1972) measure of occupational complexity, which has three dimensions: variety of occupational tasks, complexity of interpersonal relationships, and complexity of data-manipulations. The present study focuses on variety of occupational tasks.

Table 5.1 Technical, Ecological, and Social Dimensions of Four Technologies in Automobile Manufacturing[a]

Variables	Assembling Operations	Machine Operations	Test, Inspection, and Repair	Skilled-craft Operations
Control dimensions: tools and machines	Hand tools, individually controlled	Automatic machines and low control	Hand tools and high control	Semi-automatic multiple purpose machines, high control
Work flow and amount of control	Live line, low control	Hand transfer, low control	Line or hand transfer, some control	No work flow, high control
Pace of work and control	Machine determined, low control	Timed production, low control	Timed processing, some control	Untimed production, high control
Operation cycle	Invariable and short, no control	Invariable and medium, minimum control	Variable cycles possible, variable amount of control	Variable to long cycle, high control
Task differentiation	Low and relatively independent	Low and independent	Moderate and dependent	High and independent
Work attention required	Surface and brief	Surface and moderately long	Detailed and brief to long	Detailed and long
Technological interdependence	Absent	Absent	Some	Some
Technically permitted interaction	Frequent and brief	Infrequent and brief	Variable	Infrequent and long
Technically permitted cooperation	Absent	Absent	Some	Considerable
Spatial constraints and type of spatial boundaries	Small space, confined, restricted floating	Small space, confined, fixed boundaries	Variations in space, toward loose boundaries	Larger space, semi-confined, loose boundaries
Sources of influence	Extra-technical	Extra-technical	Semi-technical	Semi-technical
Communication: duration and type	Intermittent, incomplete, shouting	Intermittent, incomplete, shouting	Intermittent and variable, talking	Intermittent and complete, talking

[a] See Meissner (1969) and Goldthorpe *et al.* (1968:43-68) for related schemes.

Since the technological complexity of the factories differ somewhat, interaction rates should be expected to vary. Since the technologies associated with all work operations are simpler and more uniform in IKA and PAL than in OLDS and FIAT, the *range* in interaction permitted by technology should be smaller in IKA and PAL. Moreover, the technology of IKA and PAL should permit higher worker interaction because small and simple machines minimize spatial barriers to interaction. Departments of the four plants which are most alike in their technology should have similar rates of worker interaction. Since the machines in the parts manufacturing departments are almost identical, the interaction rates of machine operators should be most alike. Finally, complex multiple-operation machines permit their skilled operators greatest freedom to move about. Since craftsmen in OLDS and FIAT operate the largest and most complex machines, they should exhibit higher social interaction than craftsmen in IKA and PAL. In short, increasing industrialization, with its increasing technological complexity and specialization, results in conditions which produce a greater range of constraints on communication and social interaction.

A number of research findings were expected to flow from the hypothesis that both the complexity of a factory's technology and the specific technologies associated with various work operations would affect the quantity and quality of worker interaction on and off the job. The validity of such findings would be strengthened by first demonstrating that nontechnological factors such as the workers' backgrounds, seniority, job evaluations, and interpersonal relations do not influence interaction rates. If such influences are absent or can be taken into account, six major findings are anticipated on the basis of technology. First, the higher their job control, the more social contacts workers will have on the job. Second, the more complex the plant's technology, the more variation will appear in social interaction rates. Third, the less complex the factory's technology, the higher will be the population density of all departments and the more uniform will be the interaction rates. Fourth, the more social interaction on the job, the more intimate social relationships will be. Fifth, the higher the interaction at work, the more workers will make contact with each other off the job. Finally, the level of anomie workers perceive in society will not be affected by technology, but will vary inversely with the amount and quality of job interaction.

Obviously, the best data on social interaction are gathered by direct observation. Since I was not permitted to measure worker interaction directly (see Appendix B), interview questions were designed to obtain information on how far respondents could freely move while at work, how many people worked in the designated space, how many people in this area respondents could talk to, how many did the job require them to talk to, and how many did they actually talk to while working. Various ratios were constructed with this information.

Non-Technological Influences on Interaction

As suggested above, job tenure, job evaluation, background experiences, and nontechnological factors could affect the amount and quality of work contacts. All respondents had worked for the enterprises for at least one year: mean tenure in IKA was five years, the lowest of the four samples, but five years is sufficient time to stabilize social interaction rates. Three-quarters or more of the employees in each factory reported they were satisfied with daily contacts with their fellow workers and the same proportion reported satisfaction with physical working conditions. Half or more had at least two good friends in their work group and an even larger percentage reported at least one. In sum, tenure, rural-urban background, physical working conditions, job satisfaction, evaluation of fellow workers, caste (for India), and many other variables were not associated with the amount and quality of interaction on the job. If social interaction rates are associated with variations in technology, we should have considerable confidence that they are not the result of other factors.

Technological Constraint and Social Interaction

Since work operations (assembly, machining, TIR, and craft) reflect worker-machine arrangements better than skill levels do,[4] operations were used as indicators of technological constraints. Thus sweeping, assembling, or cleaning machines all represent unskilled labor performed in a wide range of technological en-

[4] The associations between type of operation and all indicators of interaction used in this chapter were higher than those between skill level and these indicators. The lower the plant's technological complexity, the more discriminating was the type of operation for interaction.

vironments, but a single operation, such as assembling, is performed in a restricted technological milieu. In automobile manufacturing, skills become increasingly complex as one moves from assembly, machine operations, TIR, to craft work.[5]

Data in Table 5.2 demonstrate that the lower the technological constraints, the more workers can move about freely on the job. This is especially noticeable at OLDS and FIAT, the factories with the most rational work organizations and the most complex technologies. The data also show that more workers in the low technological-constraint jobs report the presence of twenty-five or more work mates in their work space. As expected, this situation is most striking at OLDS and FIAT, which have the most complex technologies. Finally, the data suggest that the lower the technological constraints of work operations, the more workers talk to each other. Differences in social interaction between assemblers and machine operators were predictably small, but the differences between these two operations and TIR and the crafts were quite large.

Altogether the data in Table 5.2 support the hypothesis that the more complex the technology of the factory and the more complex the technology of the specific work operation, the less physically constrained are the workers, the more space they have to move about freely, and the more people they can talk to. This finding is borne out by the application of Goodman's test for three-way interaction effects, which is explained in Chapter 10. The distance workers can move about freely is explained more by the complexity of technology than by skill level, but the number of people found in the work space and the number of people workers can and do talk to is more highly conditioned by the skill composition of the area.[6]

[5] Unskilled workers are found among both assemblers and machine operators, and semiskilled among machine operators and in TIR. In three factories no interaction differences appeared within the skill levels for the different operations. Only in FIAT did unskilled assemblers interact more than other unskilled workers while semiskilled machine operators interacted more than other semiskilled workers.

[6] For distance a worker can move, 60 percent of the association is explained by industrialization, 29 percent by skill, and 11 percent is three-way interaction. For number of workers in the space, the percent of association explained by industry is 13; by skill, 57; by three-way interaction, 28. For number of workers talked to, the percentages are, respectively: 42, 36, and 14. All associations were statistically significant beyond the .001 level.

TABLE 5.2 SIZE AND SOCIAL COMPOSITION OF WORK SPACE FOR FOUR WORK OPERATIONS VARYING IN TECHNOLOGICAL CONSTRAINT (PERCENTS)

Technological Constraints	*Complexity of Technology* (*High*) *OLDS*	*FIAT*	*IKA*	(*Low*) *PAL*
A. PERCENT WHO CAN MOVE ABOUT FREELY AT WORK STATION				
Assembling (high)	16	12	2	77
Machining	32	10	5	78
Test, inspection, & repair	53	31	34	91
Craft (low)	79	61	38	67
Totals	47	25	16	76
Number of cases[a]	(306)	(306)	(315)	(262)
B. PERCENT WITH MORE THAN 25 WORKERS IN THEIR WORK SPACE				
Assembling (high)	13	22	24	29
Machining	16	21	15	35
Test, inspection, & repair	37	44	37	41
Craft (low)	45	82	33	28
Totals	28	38	25	32
C. PERCENT WHO TALK TO 11 OR MORE WORKERS				
Assembling (high)	28	18	8	31
Machining	28	19	1	26
Test, inspection, & repair	41	49	19	50
Craft (low)	55	79	23	37
Totals	36	37	11	33

[a] All data in the remaining tables of this chapter are based on the analytic samples and report the same number of cases as this table.

TECHNOLOGICAL COMPLEXITY, DENSITY, AND SOCIAL INTERACTION

I predicted an inverse relationship between the technological complexity of a department and its density but a direct relation between interaction rates and population density. Two indices of density were devised: the first measured the physical concentration of workers, and it was used to compare the densities directly; the second measure was adjusted for each plant to assure sufficient numbers in each density category for statistical analysis. The two measures were highly and significantly associated.[7]

[7] The corrected coefficients of contingency were: OLDS, .831; FIAT, .912; IKA, .684; and PAL, .657.

With the first index, density was inversely related to the complexity of factory technology; the more complex the technology, the lower the density (see Table 5.3). Density was lowest at

TABLE 5.3 PERCENTAGE OF WORKERS IN HIGH-DENSITY AREAS[a]

	Complexity of Technology			
	(High)			*(Low)*
Technological Constraints	*OLDS*	*FIAT*	*IKA*	*PAL*
Assembling (high)	55	59	91	94
Machining	43	51	79	93
TIR	57	75	78	96
Craft (low)	38	80	82	90
Totals	48	64	83	93

[a] Density is the number of workers in work space over work space in square yards. High density areas had a ratio over 1.0.

OLDS and FIAT, and highest at IKA and PAL. Almost all PAL workers worked in high-density areas, in contrast to about half in OLDS. As predicted, factories with the most complex technologies evidenced greatest differences in the worker density associated with the four work operations. Thus, assemblers at OLDS worked in relatively crowded quarters, and the skilled worked in spacious quarters. However, in IKA and especially in PAL, employees were equally crowded in all work operations. As anticipated, worker density in the machining departments tended to be low in all the factories because those machines were typically large. The assembly departments had higher densities. Density in the TIR departments tended to be high everywhere because the machines, tools, or gauges are small and permit crowding. But surprisingly, craft departments also showed high densities except for OLDS, which had the largest and most complex machines. Similar results to the above were obtained when I used the standardized index of density. Both indexes supported the hypothesis that the technological complexity of departments and their worker densities are related.

Whatever the technological constraints on interaction, the higher the density of the work space, the higher worker interaction should be. Table 5.4 exhibits data for workers reporting high social interaction according to variations in workspace density found in each work operation. In all plants and in each work

TABLE 5.4 HIGH INTERACTION AT WORK (TALKS TO ELEVEN OR MORE PERSONS) BY DENSITY OF WORK PLACE ACCORDING TO DEGREE OF TECHNOLOGICAL CONSTRAINT (PERCENTS)

		Complexity of Technology			
		(High)		*(Low)*	
Technological Constraints	*Density*[a]	*OLDS*	*FIAT*	*IKA*	*PAL*
Assembling (high)	Low	16	—	21	—
	Medium	26	18	32	56
	High	58	82	47	44
	Total	100	100	100[b]	100
	N	(19)	(17)	(47)	(27)
Machining	Low	38	—	26	27
	Medium	14	12	49	34
	High	48	88	26	39
	Total	100	100	101	100
	N	(21)	(17)	(39)	(56)
TIR	Low	—	—	15	—
	Medium	11	24	36	55
	High	89	76	49	46
	Total	100	100	100	100[b]
	N	(27)	(29)	(39)	(11)
Craft (low)	Low	26	—	22	11
	Medium	19	2	39	39
	High	55	98	39	50
	Total	100[b]	100	100[b]	100
	N	(42)	(48)	(36)	(28)

[a] The adjusted density index was used for this table.
[b] χ^2 not statistically significant at the .05 percent level.

operation, employees in higher-density areas tended to talk to more workers. This was especially the case in the machining and craft departments.

Social interaction should be affected by the number of persons with whom employees must communicate to do their work. In all four factories, the fewer the technological constraints on the operation, the higher the percentage of workers who reported that they had to communicate with others to do their work (see Table 5.5). For example, in OLDS only one-tenth of the assem-

TABLE 5.5 WORKERS WHO MUST TALK TO FIVE OR MORE TO PERFORM THEIR JOBS, ACCORDING TO OPERATIONS HIERARCHY (IN PERCENTS)

Technological Constraints	*Complexity of Technology* (High) *OLDS*	*FIAT*	(Low) *PAL*
Assembling (high)	11	6	20
Machining	12	12	16
TIR	21	34	27
Craft (low)	35	25	31
Totals	21	17	23

blers, compared to three-tenths of the craftsmen, had to converse with five or more persons to do their jobs. A similar ratio held for FIAT, but smaller differences were found in PAL.

Opportunities for interaction exist beyond those required by the work. In all the factories, the majority of workers needed to talk to no one or only one person, but they talked to an average of six or more. Which workers take most advantage of their interactional opportunities? To answer this, I devised a measure by dividing the number of people workers actually talked to by the number they could talk to. The percentage taking most advantage of interaction opportunities increased from OLDS to PAL (Table 5.6), following the trend of increasing densities (Table

TABLE 5.6 WORKERS WHO TOOK GREATEST ADVANTAGE OF THE OPPORTUNITY FOR SOCIAL CONTACT ON THE JOB (IN PERCENTS)

Technological Constraints	*Complexity of Technology* (High) *OLDS*	*FIAT*	*IKA*	(Low) *PAL*
Assembling (high)	17	53	45	56
Machining	15	26	41	60
TIR	12	21	43	54
Craft (low)	4	12	41	54
Totals	12	34	43	56

5.3). In all the factories except PAL, where the differences were small, assemblers took most advantage of the chance to interact. The data suggest that primitive technology is associated with high work-place density and high interaction and, as the complexity of technology increases, workers are less able to interact because the larger machines disperse them more widely.

Quality of Interaction

Social scientists have observed that social interaction on the job is often shallow, that work mates are not close friends, and that central life interests are outside the plant (Dubin, 1956; Goldthorpe *et al.*, 1968). Yet few scholars with the exception of Walker and Guest (1952) have tried to measure the intimacy or quality of social bonds among industrial workers. Since the strongest bonds should develop where the opportunities for interaction are greatest, the factors associated with high interaction (job autonomy and high density) should also be associated with strong bonds. An index of social intimacy or quality of social interaction was constructed from the following six questions (underlined responses were given a score of "one"):

> Apart from the job, how do you feel about daily contacts you have with your fellow workers? *Very satisfied or satisfied.*
>
> If you had a very important and delicate personal problem (an undesirable disease, a touchy family problem, a very large debt you could not meet, etc.) and you wanted to confide in someone, are there, among your fellow workers, people in whom you could confide and be certain that they merited your trust and would keep your secret? *Yes.*
>
> In general, how many really good friends would you say you have in your (work) group? *Two or more.*
>
> Do you exchange visits with any of your neighbors? (If yes) Are any of these your fellow workers? *Yes.*
>
> Do you get into discussions with others over economic and political issues? (If yes) With whom most often? *Mentions fellow workers.*
>
> For whose views do you have the highest regard—family, close friends, union officials, fellow workers, foremen and supervisor, political party workers, clergy, other? *Mentions fellow workers.*

All questions were associated with the total index beyond the .001 probability level as measured by the chi-square test; the uncorrected coefficients of contingency fell between .296 and .586; and the median was .465. Therefore the index is fairly reliable and, I hope, a valid measure of quality of interaction.

That half or more of the employees in the four factories scored three or higher on the six-item index suggests that moderately strong social bonds were common (see Table 5.7). Since PAL and OLDS workers had higher social intimacy scores than employees of IKA and FIAT, it is unlikely that the technological complexity of the plant or the density of the work-place accounts for the strength of bonds developed among workers. While long tenure and high job satisfaction might facilitate social intimacy on the job, in no factory were the indices of quality of interaction and job satisfaction related, and only in OLDS were tenure and quality related.

TABLE 5.7 QUALITY OF INTERACTION AMONG FELLOW WORKERS (PERCENTS)

	Complexity of Technology			
Quality Index	*(High)*			*(Low)*
Scores	*OLDS*	*FIAT*	*IKA*	*PAL*
Low: 0–1	12	27	19	11
2	28	25	30	22
3	34	25	30	32
High: 4–6	26	23	20	35
Totals	100	100	99	100

However, quality of interaction was weakly related to three indicators of quantity of interaction (see Table 5.8). The association was strongest in OLDS, where the technology was most complex, and in PAL, where departmental densities were highest. With the exception of IKA, the more workers talked to their fellow workers, the more intimate were their social ties. Except for FIAT, those who took most advantage of interactional opportunities had the strongest social bonds. Finally, a weak but consistent relationship appeared everywhere between the density of the work-place and quality of interaction. While many variables may affect the quality of social bonds which unite workers, technology appears to play a weak but persistent role. In all factories except PAL, quality of interaction was directly related to skill level and type of work operation.

TABLE 5.8 PERCENTAGE HAVING INTIMATE SOCIAL TIES WITH FELLOW WORKERS ACCORDING TO THREE INDICATORS OF INTERACTION AT WORK

Indicators of Interaction	*Complexity of Technology* (High) *OLDS*	*FIAT*	*IKA*	(Low) *PAL*
Number talked to:				
0–2	16	27	41	7
3–10	35	37	46	50
11+	48	36	13	43
Total	99	100[a]	100[a]	100
N	(79)	(70)	(63)	(92)
Possible interaction realized				
Low	3	34	2	18
Medium	31	33	55	18
High	65	33	43	63
Total	100	100[a]	100[a]	99
N	(70)	(58)	(63)	(93)
Density				
Low	31	19	27	18
Medium	20	12	37	37
High	49	70	37	45
Total	100[a]	101	101[a]	100[a]
N	(78)	(69)	(63)	(92)

[a] χ^2 not statistically significant at the 10 percent level.

Factory and Out-plant Interaction

In order to ascertain whether contacts developed on the job carry over outside the factory, an index of work-mate contacts outside the plant was developed with the following questions:[8]

Do you travel to and from work with fellow workers?

Do you exchange visits in your neighborhood with fellow workers?

Do you discuss economic and political issues with your fellow workers more than with other persons?

[8] These questions appear in abbreviated form. Positive responses were scored "one" and accumulated to form the index.

Apart from relatives, did you spend time with fellow workers during your last vacation?

Apart from the union, are your work mates members of organizations to which you belong?

Data in Table 5.9 show that the great majority of workers had no contacts or only one contact with their work mates outside the factory. Less than two-tenths can be described as having significant (three or more) social contacts. Since good comparable data are not available for other occupations, these findings are difficult to interpret. With the exception of FIAT, the more complex the technology of the factory, the higher was the amount of out-plant worker interaction. But it is more difficult to demonstrate that out-plant interaction is directly attributable to in-plant interaction.

TABLE 5.9 SCORES ON OUT-PLANT INTERACTION WITH FELLOW WORKERS (PERCENTS)

Out-plant Interaction Scores		*Complexity of Technology*			
		(High)			*(Low)*
		OLDS	*FIAT*	*IKA*	*PAL*
Low:	0	19	31	14	19
	1	34	35	34	41
	2	24	24	34	29
High:	3–5	23	9	18	11
Totals		100	100	100	100

Data in Table 5.10 reveal that in all factories except IKA, workers who talked to more fellow workers had higher interaction with them outside the factory. Also those who worked in high-density areas had more out-plant interaction with their work mates than those who worked in low-density areas. That technology is at least partly responsible for this may be apparent from data in Table 5.11. Workers in the four work operations who had high interaction with their fellow workers inside the factory were ranked for the extent of interaction with work mates outside the factory. The four items used to ascertain out-plant interaction were: meets with fellow workers in work-group outside the factory, meets with others in department outside the factory, meets fellow workers on vacations, fellow workers are

TABLE 5.10 HIGH OUT-PLANT INTERACTION WITH FELLOW WORKERS ACCORDING TO IN-PLANT INTERACTION AND WORK-PLACE DENSITY (PERCENTS)

In-plant Interaction	*Complexity of Technology* *(High)* *OLDS*	*FIAT*	*IKA*	*(Low)* *PAL*
Number talked to:				
0–2	19	25	48	11
3–10	40	29	36	49
11+	41	46	16	41
Total	100[a]	100	100[a]	100
N	(69)	(28)	(56)	(105)
Density Index:				
Low	32	29	29	21
Medium	19	7	37	34
High	49	64	34	45
Total	100	100	100[a]	100[a]
N	(68)	(28)	(56)	(105)

[a] χ^2 not statistically significant at the 10 percent level.

members of same organization exclusive of the union. In all four factories, assemblers and machine operators had less out-plant interaction with their fellow workers than those in TIR and the crafts. The differences among the four work operations were greater for OLDS and FIAT than for IKA and PAL. Again, the more complex the technology, the greater the differences among

TABLE 5.11 CUMULATIVE RANKS FOR FOUR QUESTIONS ON WORK-MATE INTERACTION OUTSIDE THE FACTORY FOR WORKERS WHO HAD HIGH INTERACTION IN THE FACTORY

Technological Constraints	*Complexity of Technology* *(High)* *OLDS*	*FIAT*	*IKA*	*(Low)* *PAL*
Assembling (high)	6.0	6.5	8.0	7.5
Machining	6.5	5.5	4.0	8.5
TIR	13.5	13.0	13.0	14.5
Craft (low)	14.0	15.0	15.0	9.5

the work operations for the amount of outside interaction workers had with their work mates.

Interaction and Anomie

Finally, I hypothesized that the anomie which workers perceive in society is not related to the plant's technological complexity but is inversely related to the amount and quality of contacts with fellow workers. Since increasing technological complexity increases interaction in some circumstances and decreases it in others, its overall effects might be indeterminate. However, workers with the most and strongest ties with their fellow workers on and off the job, regardless of the technology of their plants, should be most socially integrated and should perceive least societal anomie. A five-item Guttman scale of anomie (normlessness) was constructed which had coefficients of reproducibility above .85 in each country (see Chapter 11).

As expected, the plant's technological complexity and anomie scores were not unrelated. The Indians had the highest median scores (4.7), followed by the Italians (3.7), Americans (2.9), and Argentinians (2.0). Surprisingly, no statistically significant associations appeared among any of the eleven indicators of quantity or quality of work interaction and anomie scores. Occasional weak associations did not conform to any pattern.

Conclusions

I hypothesized that the quantity and quality of social interaction among workers is conditioned by the plant's technological complexity and the constraints imposed by work operations. The modern factories (OLDS and FIAT) with the largest machines dispersed workers and decreased their interaction. The more workers controlled their machines and the more crowded their work stations, the more they interacted with others and the more they took advantage of interactional opportunities. Differences in interaction rates among the various operations were largest in the plants with the most complex technologies. The more specialized the technologies, the more distinctive was their impact on interaction. Although the evidence was not strong, quantity and quality of social interaction appeared to be related. In addition, the more workers interacted in the factory, the more they saw each other in the outside community. However, the amount of interaction on or off the job did not affect the worker's perceptions of societal anomie. Why was this?

Although some students (Blauner, 1964) relate anomie to the dehumanizing conditions of industrial work, typical anomie scales tap feelings of optimism. Tudor (1972) found that job complexity was not related to feelings of powerlessness among American men, and Seeman (1971:395-396) found no relation between work-related beliefs of Parisian workers and their feelings about national events. Apparently workers can be socially integrated at work and yet feel powerless about events in society. This study found that the quantity and quality of work interaction in all four plants was relatively high and that most workers were satisfied with their jobs and their contacts with their workmates. Contrary to popular belief, the factory may create integrated social systems rather than anomic ones.

The evidence of this chapter supports the view that the social life of the work situation is strongly conditioned by technology. Even where industry is in its infancy, factory social life is similar to that found in mature industrial societies. It appears, therefore, that current explanations of worker behavior need to be reevaluated. The human relations view that management is responsible for the social environment is too simplistic, as is the humanist view which sees the factory as antithetical to social well-being. Finally, the traditional Marxist view that the rationalization of production inevitably leads to worker dissatisfaction and rebellion against the political order seems too simple and in need of reexamination (Seeman, 1971).

6

Autoworkers and Their Machines

Intellectuals hate machines and have elaborated an anti-machine ideology (Report, 1973). While they admit in passing that machines have lightened the burdens of labor, they describe in great detail how machines have unnaturally intruded upon the lives of workers. Often reconstructing a past golden era where all males were skilled, intellectuals emphasize that workers once joyfully practiced their crafts in the warm bosoms of their families. But machines gradually destroyed their crafts and today only a few skilled workers survive in industry. Inexorably machines are robbing workers of their skills, isolating them from one another, and creating alienating work environments. When workers finally recognize what is happening to them, they will reverse the order of things and subjugate machines to their needs.

In spite of prevalent ideology, countless studies report that most industrial employees are satisfied with their work (Blauner, 1960:341). Researchers who normally accept survey data often reject the findings on work satisfaction; they redesign or reinterpret studies to prove that factory employees are dissatisfied. Automobile assembly workers are selected as the classic case of dissatisfaction because their operations are most subdivided, most repetitive, most meaningless, and most subject to machine control (Chinoy, 1955; Blauner, 1964; Kornhauser, 1965; Sheppard and Herrick, 1972).

The conflict between the ideology and research findings on work satisfaction raises certain fundamental questions. Do industrial workers hate the factory and their jobs? Does this hatred increase with industrialization? Do workers at different skill levels respond differently to industrialization? This chapter attacks these questions in three areas: work satisfaction, satisfaction with factory employment, and satisfaction with specific physical and social job conditions. I made one major assumption: what employees say about their work is as valid as anything they say about themselves.

I expected employees in all four factories, irrespective of their skill or the complexity of local technology, to be generally satisfied with factory employment and their jobs. The extent of satisfaction in the three areas would increase according to the workers' skill level and their control over work operations (assembly, machining, test-inspection-repair, and craft work). Satisfaction would also vary inversely with the technological complexity of the factory: the more complex, the less the satisfaction.

Industrial Development and Worker Satisfaction

Work

A rationale is needed to explain how industrial growth affects satisfaction with work, factory employment, and specific job routines. The Protestant Ethic is often invoked to explain the origin of work attitudes in western societies. While the Ethic may have motivated entrepreneurs to accumulate capital, it probably had little effect on the work values of early industrial employees. With some exceptions (Kuczynski, 1971), scholars of industrialization (cf. Tilgher, 1930; Ellul, 1967; Caplow, 1954; Berger, 1964) have neglected the subject of how early industrial employees felt about their work. Many scholars mistakenly assume that artisans typically become factory employees[1] (Bauman, 1972), but early industrial employees were mostly women, children, and ex-peasants who were too preoccupied with survival to worry about work satisfaction. Work was simply part of living. Moreover, the joyful work of the post-Renaissance period is probably a myth (de Man, 1929:146). Not all villagers were artisans, and many who worked in the shops performed dull and routine work which differed little from that in the factory.

Conditions in contemporary industrializing societies resemble the post-Renaissance era in Europe in some respects. Feldman and Moore (1960:41-61), Lambert (1963:180), Morris, (1960:189), and Blumer (1960) report that where poverty is widespread, workers are primarily concerned with survival, regularity of income, and job security. The belief that work should be personally fulfilling probably arose after the market became fully monetized, after consumption levels rose above that of subsis-

[1] In newly industrializing countries, craft workers cling to their work, while relatively inexperienced rural and urban workers move into the factory directly. Craft work in fact is never completely phased out (Scoville, 1974).

tence, after job choice was possible, and when workers became geographically and occupationally mobile. To generalize: the less industrial a society, the less salient is work satisfaction for factory employees; the more industrial, the more workers feel that work should be satisfying. OLDS, FIAT, and IKA respondents live in societies where work-satisfaction norms are operative. Since the conditions for developing such norms are limited in India, PAL's employees should be least concerned about work satisfaction.

The Factory as a Work Locus

Popular writers assume that most workers dislike the machine environment of the factory and prefer either the natural environment of agriculture or the human environment of the service sector. Factory work presumably restricts the freedom and autonomy which ex-farmers earlier enjoyed (Balandier and Mercier, 1962), while urbanites hate the anonymity of their lives and yearn for the human contact found in the service sector. To my knowledge, all this is speculation; there is little theory or research on the subject of how workers respond to the changing physical and social environment of the agricultural, manufacturing or service sectors (Clark, 1957).

I hypothesize that the sector which offers workers the greatest economic security and opportunity for mobility is the sector they prefer. Where industry is marginal, pays low wages, and offers irregular employment, workers probably define farming as more satisfying (cf. Moore, 1951:61-66). Gregory (1960:144-150) reported that agricultural workers in Puerto Rico rationally calculated the economic advantages of becoming industrial employees; when the advantages appeared to favor industry, they anticipated liking factory work. Where manufacturing is growing rapidly and provides high wages and opportunity for upward mobility, as in Italy and Argentina, both rural- and urban-born workers probably define factory employment as more satisfying than other work. In the United States, where the service sector provides most opportunities, workers define it as most attractive.

Since opportunities vary by skill, workers with different skills may evaluate factory employment differently. As aristocrats of labor, skilled workers have most opportunities and probably see factory employment as most desirable. Since the skilled were in shortest supply in the rapidly industrializing countries of Italy and Argentina, I anticipated that they would be more satisfied with factory work than the skilled in the United States and India. Economic opportunities among the less skilled probably vary with the extent of national industrialization. In countries beginning to industrialize, unskilled factory workers are relatively privileged economically and they probably prefer factory work more than do their counterparts in industrial nations. However, workers everywhere quickly learn that the most rewarding and prestigious jobs are found in the service sector (Inkeles and Rossi, 1956), so everywhere they hope their children will move into such jobs.

Intrinsic Job Satisfaction

Although Inkeles (1960) observed that factory workers in many countries are satisfied with their jobs, no one has explained how satisfaction with different physical and social attributes of their jobs changes with industrial growth. An explanation should consider such factors as the prevalence of manufacturing, the pace of industrialization, the scarcity of skills, and the complexity of industrial technology. In new industrial nations, job satisfaction probably has low salience for most workers because holding a factory job is relatively prestigious (cf. Germani, 1966). Recruited from agriculture, most new industrial employees are accustomed to exhausting toil. To them industrial jobs pay relatively well, are physically undemanding, and therefore satisfying. Where manufacturing is expanding rapidly and opportunities for job advancement are high, workers are better satisfied with their jobs than where opportunities are stable or declining, as in the United States.

Skilled workers everywhere are the most satisfied with all aspects of their jobs (Inkeles, 1960:15-23). When employers fail to utilize their skills and when craftsmen reach their wage ceilings quickly, as in Italy and Argentina, they may become dissatisfied. Rural-born semiskilled workers experience most upward mobility where industry is just beginning, and they probably are highly satisfied with their jobs. In automobile manufacturing, the ratio of unskilled to skilled is highest where technology is most complex, as in the United States. Overeducated for their jobs, American unskilled workers are probably more dissatisfied with their jobs than are similar workers in less developed economies where unskilled factory employees get more money and prestige than most manual workers. The less industrialized the society, the more the unskilled stress the economic rather than psychic rewards of jobs. Occasionally, skilled craftsmen must decide whether to stay with a satisfying job with low pay or take a routine job which pays more. Goldthorpe *et al.* (1968:33) and Walker and Guest (1952:64) found that they usually select routine jobs which pay more.

Many studies have shown that workers who have the least control over their machines are most dissatisfied with their jobs (Blauner, 1960:345-349; Friedmann, 1955:129-156). In this study, assemblers had the least job control, followed by machine operators, those in test-inspection-repair, and craft workers (cf. Wyatt and Marriott, 1956). Assemblers should be the most dissatisfied. Assembly lines in OLDS and FIAT were equally rationalized, but OLDS's line moved faster; IKA rotated assemblers from one slow-moving line to another; and PAL's assemblers, who used hand tools almost exclusively, worked at the most leisurely pace. Differences in the organization of production suggest that OLDS assemblers should be the most dissatisfied, followed by FIAT, IKA, and PAL assemblers. Even though assembly jobs are monotonous, I predicted that assemblers in all countries would be satisfied with their jobs, but would complain when the work-pace was suddenly changed and/or increased.

Finally, job satisfaction tends to increase with extent of integration into the work-group, union, and community (Friedmann, 1955:291-400). Although group cohesiveness increases with the freedom to communicate on the job (Meissner, 1969; Sayles, 1963), job satisfaction and freedom to communicate do not necessarily go together. Skilled workers, who have the most opportunity to communicate, probably attach the least importance to it because the work itself, its status, and the income are satisfying. Since automobile technology limits the interaction of unskilled workers the most (see Chapter 5), they probably attach the most importance to the ability to move and communicate on the job.

Salience of Work Satisfaction

According to the theory developed above, work satisfaction should be least salient to unskilled workers in newly industrializing societies. When asked whether people should expect satisfaction from working, almost everyone in OLDS and IKA agreed that they should.[2] But 5 percent of FIAT and 18 percent of PAL employees did not; they felt that good pay and economic security were sufficient rewards for working. Among PAL employees who did not endorse the work-satisfaction norm, half could give no reasons for their position. Workers who endorsed the norm in the other plants readily provided reasons for their beliefs, but one-quarter of PAL's employees gave vague and irrelevant replies. Altogether, almost half of the Indian respondents either did not support the norm of work satisfaction or could give no reasons for supporting it.

The responses of OLDS and IKA employees point to the salience of work satisfaction in mature industrial societies, the responses in FIAT's to an intermediate pattern, and those in PAL to an early industrial pattern (see Table 6.1). More than half the

TABLE 6.1 REASONS WHY SATISFACTION SHOULD BE EXPECTED FROM WORK (PERCENTS)

Reasons	*OLDS*	*FIAT*	*IKA*	*PAL*
Motivates working	39	20	40[a]	22
Enhances work creativity and meaning	29	13	16	0
Makes job tolerable	28	29	26	1
Gives life order and meaning	—	36	11	25
Other responses, vague responses, don't know	4	1	7	52
Total	100	99	100	100

[a] Includes 11 percent who believed that satisfaction motivates occupational mobility.

workers in OLDS and IKA and a third in FIAT thought that work satisfaction was needed to motivate work or to enhance its meaning, but less than one-quarter of PAL employees related work satisfaction to work itself. Except for PAL, about three-tenths of the respondents in each plant felt that job satisfaction served to reduce the feelings of monotony on the job. Briefly, these data underscore the lower salience of job satisfaction in India where one-half the respondents could provide no reasons why a job should be satisfying.

[2] Data from the representative sample are given for descriptive purposes, but when skill level and work operation are used as independent variables, the expanded samples for IKA and OLDS are used.

Walker and Guest (1952); Chinoy (1955); Friedmann, (1955); Blauner (1964), and many others have described the boredom and alienation of assembly-line workers. But Dubin (1956) found that since industrial workers do not consider work to be a central life interest, they suffer little from job dissatisfaction. Goldthorpe *et al.* (1968:53) also found that English automobile workers have an instrumental view of their work. It is a means of providing an income to support a desired level of living; enjoyment is secondary.

Following a lead from Morse and Weiss (1955), respondents were asked: "If, without working, you were guaranteed an income equal to your present wages, would you continue to work at your present occupation? This means you couldn't accept another paying job." Morse and Weiss (1955:197) found that three-fourths of working-class respondents would continue working, but only one-third would continue in the same occupation. The question in this study gave respondents no choice; they had to continue in their present occupations. Surprisingly, a higher percentage of autoworkers in the four factories were attached to their work than were the workers in the U.S. sample Morse and Weiss studied: half or more in contrast to one-third would continue working in their present jobs even if given the choice of leisure (see Table 6.2). While Morse and Weiss (1955:197) found that more of the skilled than the less skilled were attached to their work, in this study as many assemblers were attached as craftsmen.

With the exception of IKA, the reason most often given for preferring work to leisure was that work prevents boredom; two-fifths to three-quarters could think of no other activity to replace work. The more industrial the country, the more this reason was given. In support of the theory, more workers in the rapidly industrializing countries (Argentina and Italy) preferred work to leisure because they liked their jobs or wanted to be promoted. Probably because IKA workers were younger than the others, they were the most highly work-motivated (cf. Briones, 1963:579). One-third of PAL's workers, but very few in the other factories, felt that they could not accept wages without working. Two possible interpretations are suggested: the norms of reciprocity are so highly developed in India that wages cannot be accepted without a service being given in return, or the idea of receiving wages without working is so absurd as to be rejected outright.

TABLE 6.2 PERCENTAGE FOR WHOM WORKING SERVES NON-MONETARY FUNCTIONS AND REASONS GIVEN FOR WORKING OR NOT WORKING

Reasons	*OLDS*	*FIAT*	*IKA*	*PAL*
Would continue working	64	51	51	90
N	(160)	(156)	(133)	(234)
Reasons for continuing:				
Activity: keeps busy, prevents boredom	76	49	18	37
Life: gives direction, meaning, purpose to life	8	8	5	16
Job: likes work or wants promotion	9	28	63	11
Moral: couldn't take wages without working	4	—	6	32
Other reasons	2	14[a]	8	4
Total	99	99	100	100
N	(160)	(156)	(133)	(234)
Reasons for not continuing:				
Prefers other free-time activities to work	52	40	25	4
Doesn't like job or work	11	33	39	81
Doesn't like factory work	14	17	26	—
Other reasons[b]	23	11	10	15
Total	100	101	100	100
N	(88)	(150)	(128)	(27)

[a] Current income not sufficient to make the choice possible.
[b] Largely reasons of health and approaching retirement.

A minority of employees in OLDS and PAL and about half in FIAT and IKA preferred other activities to working (see Table 6.2). The less industry in the country, the more workers preferred leisure because they disliked either working, their jobs, or the factory. Predictably, the reasons for wanting to work varied with job operations. Assemblers, especially in OLDS but also in FIAT, would continue working because work organized their lives; the skilled, because they were involved in the work itself. The failure of these trends to appear with other samples means that the theory is only partly supported.

In summary, although the norm of work satisfaction was the least salient to workers in the least industrialized societies, the norm's primary function was seen as motivating work and preventing boredom. Contrary to a widespread view, most employees preferred work to idleness because they could conceive of no satisfactory substitute activity (see Friedlander, 1966). Most support for the work ethic was found in countries where industry was expanding rapidly, while alternative patterns to work were more popular in the more industrial societies. To the extent that workers were aware of their own motives, these findings fail to support Vroom's contention (1964:30-41) that interaction at work and the social status of a job motivate working.

Evaluation of Factory Employment

Contrary to the anti-industrial ideology, I anticipated that workers would prefer the factory to agricultural or service employment and that this preference would increase with skill level. To eliminate the influence of income, the question was asked: "With the same hourly wage, would you prefer to work on a farm, an office, or a factory machine?" Data in Table 6.3 show

Table 6.3 With the Same Hourly Pay, Would You Prefer to Work on a Farm Machine, an Office Machine, or a Factory Machine? (percents)

Sector Preferred	*OLDS*	*FIAT*	*IKA*	*PAL*[a]
Farm	41	22	7	53
Office	10	25	22	9
Factory	49	53	71	38
Total	100	100	100	100
N	(241)	(295)	(274)	(252)

[a] Question for PAL: Which occupation is most desirable: office work, skilled factory work, or independent farming?

that a majority in all factories except PAL preferred the manufacturing sector.[3] Argentinians, most of whom were reared in cities, least preferred farming while Americans and Indians least

[3] Data for PAL are not strictly comparable. Given the low technological state of Indian agriculture, it seemed unrealistic to inquire about work on agricultural machines. Therefore, a question was substituted which asked workers to select the most desirable occupation: office worker, skilled factory worker, or independent farmer.

preferred the white-collar sector. More of FIAT's and IKA's employees were attracted to the office sector than OLDS and PAL workers. Except for PAL, the theory was supported: the less industry in a country, the more workers preferred the factory to farming; the higher their skills, the more workers preferred the factory.

Respondents were then asked to rank three occupations which represented the sectors (office worker, skilled factory worker, and independent farmer) along five dimensions: monotony, desirability, prestige, necessity, and capacity to provide work satisfaction (see Table 6.4). The data support earlier findings. The factory was most attractive in Italy and Argentina where industry was expanding most rapidly; workers ranked the factory as most desirable, most necessary, least monotonous, and most satisfactory. Factory employment was least appealing where industrialism was most highly and least highly developed: United States and India. OLDS and PAL employees ranked agricultural work as most desirable, most necessary, most satisfactory, and least montonous. While workers generally recognized the high prestige (respect) of white-collar employment, they thought it was the least necessary, least desirable, most monotonous,[4] and least satisfying.

Three indexes of sector attractiveness were devised by adding the ranks given each occupation on the five dimensions. Office work was least attractive, except for FIAT; farming appealed more to workers in PAL and OLDS; and IKA and FIAT employees found the factory most attractive and the farm least attractive (see Table 6.4).

How can the differences in sector attractiveness be explained? In general, the data support the theory that the more rural the society, the more factory work is preferred. Why PAL deviated from the trend is puzzling. Perhaps the workers were genuinely rurally oriented; relatively more of PAL's employees had worked in agriculture and many still had relatives living in the villages. Some could still consider farming as a realistic alternative to factory work because they had economic and kinship ties to their villages.[5] Yet their attachment to agriculture was clearly transi-

[4] Office work was ranked as more monotonous than factory work in all four plants, especially by skilled workers.

[5] Lambert (1963:83) reported that 31 percent of his urban-born sample of industrial workers in Poona wanted to "round out their lives in a village."

TABLE 6.4 EVALUATION OF FIVE ASPECTS OF OCCUPATIONS REPRESENTING THREE SECTORS (PERCENTS) AND MEAN RANKS FOR SECTORS

Dimension of Occupations	*OLDS*	*FIAT*	*IKA*	*PAL*
Occupation most desirable:				
Small independent farmer	44	16	28	53
Skilled factory worker	39	44	56	38
Office worker	14	37	15	9
Other	3	3	1	–
Occupation most respected:				
Small independent farmer	27	6	18	48
Skilled factory worker	12	27	43	21
Office worker	39	53[b]	36	31
Other	22[a]	14[b]	3	–
Occupation most necessary:				
Small independent farmer	59	32	40	77
Skilled factory worker	14	39	50	21
Office worker	3	8	5	2
Other	24[a]	21[b]	5	–
Occupation which gives most satisfaction:				
Small independent farmer	65	21	29	57
Skilled factory worker	20	56	59	32
Office worker	5	18	9	10
Other	10	5	3	1
Occupation which is most monotonous:				
Small independent farmer	9	21	20	17
Skilled factory worker	34	21	18	12
Office worker	51	51	59	69
Other	6	7	3	2
Index of industrial sector attractiveness	1.6	2.2	2.9	2.0
Index of office sector attractiveness	1.2	1.6	1.2	1.2
Index of agricultural sector attractiveness	2.0	1.0	1.4	2.5

[a] Mostly "don't know."
[b] Mostly "all the same."

tional because, like workers elsewhere, they did not want their children to become farmers.

Almost none of the OLDS workers, despite their small-town socialization, had worked in agriculture, and most of their relatives had left farming. In Chinoy's (1955:119) earlier study of

OLDS in the late forties, almost half the workers thought about starting their own businesses, and almost a fifth about owning farms. But in this study, only one-tenth had ever thought about farming and fewer considered business. Over the years OLDS workers may have become more realistic, probably concluding that they could not accumulate the capital needed to buy farms or businesses. Although a larger percentage of FIAT than OLDS employees had earlier worked in agriculture, the Italians strongly rejected the idea of returning to the farm, as did the Argentinians, most of whom were reared in cities. More FIAT and IKA respondents than OLDS and PAL were attracted to white-collar jobs as a second choice. In all four factories more of the skilled than the less skilled preferred factory to other employment. Finally, whatever their sector preferences, very few workers objected to the factory as physically confining.

In sum, workers everywhere chose the factory over the office. The Indians preferred farming over the factory, the Italians and Argentinians preferred the factory, and the Americans, while recognizing the prestige of the office and the desirability of farming, settled for factory work as second-best.

Intrinsic Job Satisfaction

The decision to stay with factory work is undoubtedly influenced by the response to the job itself. But job satisfaction is not a simple response to the work; it is influenced by the age of the employee, his marital responsibilities, socialization, expectations for mobility, experience with unemployment, how far he must travel to work, the effectiveness of the union, and many other factors. However, I found that none of these conditions was associated with job satisfaction in any country, with the exception that older second-generation factory employees in IKA were more satisfied than others. These negative findings suggest that attributes of the job itself were critical in determining satisfaction.

Table 6.5 presents data for eighteen indicators of job satisfaction. In conformity with the findings of other studies (Gurin *et al.*, 1960:163), most employees reported satisfaction with most aspects of their jobs. Seven- to nine-tenths, in all plants, replied affirmatively to direct questions about their job satisfaction. Almost all workers reported satisfaction with the physical locale of their work and their daily contacts with fellow workers. IKA's

TABLE 6.5 DIMENSIONS OF JOB SATISFACTION (PERCENTS)

Dimensions	*OLDS*	*FIAT*	*IKA*	*PAL*
Satisfied with job—with work routines	82	76	88	68
Satisfied with general occupation	81	77	89	65
Satisfied with location of work	80	73	76	74
Satisfied with daily contacts with fellow workers	91	86	94	74
Rates occupation as good or very good	68	68	89	59
Likes present job more than others	58	48	61	55
Not planning a job change	77	83	65	N.A.
Would not like to change type of work activity	52	52	26	55
More satisfied with job than fellow workers	35	4	16	41
Family rates job as good or very good	65	68	95	55
Friends rate job as good or very good	65	70	94	63
Dislikes nothing about present job	18	N.A.	32	60
Never thought of having another job	33	46	54	N.A.
Fellow workers are satisfied with their jobs	63	48	80	55
Wants to change work routines less than once a month	39	67	43	70
Index of overall work satisfaction	4.5	4.5	4.8	4.6[a]
Index of negative aspects of job satisfaction	2.8	2.4	2.7	2.3
Index of reference group job evaluation	3.7	4.2	4.4	3.4

[a] Overestimate because different question was used.

employees were the most satisfied, followed by OLDS, FIAT, and PAL. Thus, no support was found for the hypothesis that job satisfaction is inversely related to industrial development.

The first eleven indicators of job satisfaction in Table 6.5 were analyzed according to skill level and work operations. As expected, in all factories the higher the worker's skill level and the more he controlled the machine, the more satisfied he was with his job. The more industrial the country, the stronger the trend. Job satisfaction was even more highly associated with the worker's operations (assembly, machining, etc.) than with his skill level. In each factory, nine of the eleven indicators of job satisfaction were associated with work operation; the more workers controlled operations and the more complex their machines, the more they were satisfied. Identical results appeared for job evaluation; workers with highest skill and greatest control over work operations rated their jobs highest (see Table 6.6).

These findings generally support the theory I proposed. In each society, high skill and high work control were associated with job satisfaction. Although the trend was weaker in the

TABLE 6.6 JOB SATISFACTION AND JOB SELF-RATING OF WORKERS BY SKILL LEVEL AND OPERATIONS PERFORMED (PERCENTS)

	OLDS	*FIAT*	*IKA*	*PAL*
Skill Level:		Satisfied with Job		
Unskilled	71	68	86	58
Semiskilled	85	75	84	74
Skilled	93	93	96	75
Total	84	76	87	68
Operations:				
Assembly	75	71	86	65
Machine operator	82	71	81	62
TIR	85	76	89	50
Craft	92	92	97	70
Total	84	76	87	64
Skill Level:		Job Rated Good or Very Good[a]		
Unskilled	53	41	84	53
Semiskilled	67	51	91	58
Skilled	94	70	99	69
Total	71	52	90	59
Operations:				
Assembly	51	40	87	50
Machine operator	63	44	86	60
TIR	72	62	89	59
Craft	90	72	100	68
Total	71	52	90	59

[a] Categories for OLDS and IKA were "very good" and "good," for FIAT and PAL, category "very good" was not in the interview.

less mechanized plants, it held both in IKA, where job satisfaction was highest, and in PAL, where it was lowest. As predicted, workers with lower skills were *relatively* more satisfied in the less industrialized countries. Where industry is new, perhaps even unskilled factory workers feel that they are comparatively skilled (Froomkin and Jaffe, 1953) or where unemployment is high, the unskilled may feel lucky just to have a regular factory job (cf. Lambert, 1963:35).

Indirect Approaches to Job Satisfaction

Some students (Sheppard and Herrick, 1972:54) say that direct questions on job satisfaction evoke favorable responses.

Therefore, workers were asked three indirect questions: what job they most preferred among all those held, whether they contemplated a job change, and what job they aspired to for themselves and their children. The selection of the present job over others should reveal some degree of job satisfaction. As predicted, the majority of employees in all plants except FIAT preferred the present job over all previous jobs. I predicted that this would hold most in new industrial societies where the automobile industry pays relatively high wages, but no national differences were found (see Table 6.7). Parenthetically, about half the respondents indicated present job preference on the grounds that it was the most interesting. This finding challenges the view that autoworkers have a purely instrumental view of their jobs (Goldthorpe *et al.*, 1968).

TABLE 6.7 JOB LIKED MOST AND REASONS FOR CHOICE ACCORDING TO SKILL LEVEL AND OPERATION PERFORMED (PERCENTS)

Job Preferences	*OLDS*		*FIAT*		*IKA*		*PAL*	
Present job liked most	58		48		61		55	
Reasons why liked most:								
Job interest	52		55		47		54	
Income	27		26		27		27	
Other	21		19		26		19	
	Reasons Why Job Liked Most[a]							
Skill and operation	*Interest*	*Income*	*Interest*	*Income*	*Interest*	*Income*	*Interest*	*Income*
Skill level:								
Unskilled	42	39	50	30	39	30	52	19
Semiskilled	56	33	64	11	38	24	52	23
Skilled	65	25	77	18	41	25	59	25
Total	55	33	55	26	39	27	54	27
Operations:								
Assembly	49	16	49	30	42	30	49	30
Machine	43	22	50	26	40	7	67	17
TIR	54	28	57	17	42	33	48	37
Craft	71	12	70	13	60	20	64	26
Total	54	20	53	24	44	22	58	27

[a] Based on analytic sample.

I expected skilled workers to consider job interest to be the most important basis for job preference, and the unskilled to rank income highest, especially in the less industrialized societies. Data in Table 6.7 show that differences in job interest according to skill and work operation were greatest in the most industrialized countries (OLDS and FIAT). The inference may be drawn that the fewer references to job interest by the skilled in IKA and PAL means that they were doing more routine work. Predictably, assembly-line workers liked their jobs more for economic reasons than for interest, in FIAT and IKA (cf. Tausky, 1969), but not in OLDS and PAL. We cannot conclude from these data that people in routine and monotonous jobs prefer them because of their wages.

If a job change is persistently on the worker's mind, he probably is unhappy with his present job; about one-quarter in each country were so dissatisfied (Table 6.8). Surprisingly, about one-fifth of the OLDS workers who thought about other occupations (almost none of the others) named farming as an alternative job, but about two-fifths in each plant named occupations in the industrial sector, and one-fifth specified professional jobs. More workers in the less developed societies named managerial and business jobs. Whatever their thoughts about other occupations, the overwhelming majority (85 percent) were not planning to quit their present jobs, but those who did were pessimistic about their chances of finding desirable work.

The data from this study show weak or inconsistent relationships between the worker's skill level or activities and his plans to seek other jobs. Assemblers were not planning to leave the factory any more than other workers. Among those desiring to change jobs, the unskilled wanted more skilled factory jobs and the skilled, white-collar, and proprietary jobs. But skilled workers were not more optimistic than assembly workers about their chances of finding good jobs.

Occupational bliss is indicated when workers want their children to inherit their occupations. Only 10 percent of the respondents wanted their children to be in any type of manual work, but more than two-thirds hoped that their children would become professionals. Despite the great personal appeal which business, the crafts, and farming had for many workers, few wanted their children in these jobs. Of those who wanted their children to become professionals, half expected them to achieve

TABLE 6.8 PERCEIVED OCCUPATIONAL ALTERNATIVES FOR AUTOWORKERS (PERCENTS)

Question	*OLDS*	*FIAT*[a]	*IKA*
Thought of having a different occupation?			
Very rarely	33	46	54
Sometimes	44	28	21
Often	23	26	25
Total	100	100	100
What occupation have you thought of?			
Farmer	18	3	3
Unskilled and semiskilled manual work	5	31	18
Skilled manual work	40	20	24
White collar and professional	19	19	20
Proprietor or manager	18	26	35
Total	100	99	100
Planning a job change? (Yes)	15	17	30
N	(37)	(51)	(82)
What are your chances of getting that job?			
Good	26	16	37
Fair	15	74	27
Poor	59	10	37
Total	100	100	101

[a] For FIAT, the order of the questions was different; workers planning a job change were asked about their chances, whereas the others were asked about their chances after having named the occupation.

the goal. Only the OLDS sample had a sufficient number of older workers with enough employed children to describe their employment with confidence. Achievement levels were quite high; while six-tenths of the fathers expected their children to become professionals or managers, three-tenths were in these occupations, one-tenth were skilled workers, and the remaining six-tenths were unskilled and semiskilled manual workers.

A common stratification principle is that people in prestigious jobs have high occupational aspirations for their children and succeed in placing them in high-status jobs (Blau and Duncan, 1967; Sewell *et al.*, 1970). The principle holds even among manual workers. More skilled than unskilled workers in each sample wanted and expected their children to become professional workers. Only for OLDS was it possible to examine the effect of stratal origin of parents on the occupational attainment of children. As expected, more children of skilled than of less skilled employees became skilled or professional workers.

In brief, regardless of national economic development, industrial workers everywhere know the occupational structure (Hodge *et al.*, 1966) and want their children to achieve the highest levels. Occupational expectations for their children roughly parallel the workers' own positions in the skill hierarchy. The fact that workers had lower occupational aspirations for themselves than for their children does not lead to the conclusion that workers were dissatisfied with their own jobs. On the contrary, the data suggest a universal principle: that industrial workers are relatively content with their own status but they do not expect their children to be in the working class.

Monotony and Job Dissatisfaction

The extensive literature on worker discontent focuses on the unskilled, who are condemned to monotonous jobs. Social psychologists (Faunce, 1968; Kornhauser, 1965; Sheppard and Herrick, 1972) insist that the absence of work autonomy causes unhappiness, anomie, and alienation. To assuage their suffering, workers are said to demand higher wages and improved working conditions (Goldthorpe *et al.*, 1968). But evidence that the auto assembly worker is suffocated by monotony is far from conclusive (Clack, 1967). Kornhauser (1965:158), among others, feels that workers do not admit unhappiness with their work because such an admission acknowledges failure in a key life role. Therefore, valid data on job dissatisfaction must be obtained indirectly by comparing feelings about current and previous jobs and by encouraging workers to explore the negative aspects of their jobs. I used this indirect approach, but expected to confirm earlier findings that job discontent is limited but increases with industrialization.

After exploring the extent of their job satisfaction, I asked the workers for the reasons for their feelings. Job interest, good working conditions, and type of work were mentioned as the main sources of job gratification, but the joys of social interaction received almost no recognition. In the four factories, less than 10 percent mentioned monotony. OLDS and FIAT employees were equally concerned with poor working conditions, while the FIAT and PAL respondents pointed to wage and seniority problems as causes of dissatisfaction (see Table 6.9).

TABLE 6.9 REASONS FOR FINDING JOB SATISFYING OR DISSATISFYING (PERCENTS)

Reasons	*OLDS*	*FIAT*	*IKA*	*PAL*
NEGATIVE REASONS:				
Monotony	7	7	5	10
Poor working conditions	5	7	2	3
Other	4	10[a]	8	18[c]
POSITIVE REASONS:				
Job interest and job control	24	23	15	25
Good working conditions	10	13	29	14
Likes type of work, used to it	29	15	26	21
Prestige or recognition of job	5	6	12	—
Other	16[b]	19[d]	3	8
Total	99	101	100	99

[a] Six percent were "resigned" without reason given.
[b] Eight percent indicated inadequate pay and 7 percent insufficient seniority.
[c] Nine percent gave ambivalent responses.
[d] Seventeen percent ambivalent, vague, unclassifiable, responses.

Job satisfaction was then analyzed by skill level and type of work activity. The expectation that the unskilled and assemblers would complain most about monotony was supported in FIAT and PAL, but in OLDS and IKA they were most concerned about good working conditions (space, light, noise, ventilation).

Blauner (1960:355) and others claim that workers follow a norm of defining their jobs as satisfactory and that only a bold minority dares to expose the general unhappiness. Respondents were therefore asked directly what they specifically disliked about their jobs. The most frequent reply was "nothing," a response given by 60 percent of PAL's employees, 35 percent of IKA's, and 18 percent of OLDS' (see Table 6.10). Monotony and lack of job control were not major complaints;[6] more complained

[6] Goldthorpe *et al.* (1968:18) reported that 67 percent of assemblers replied affirmatively to the question, "Do you find your present job monotonous?"

TABLE 6.10 RESPONSES TO QUESTION: WHAT DO YOU MOST DISLIKE ABOUT YOUR JOB? (PERCENTS)

Job Attributes	*OLDS*	*IKA*	*PAL*
Nothing	18	35	60
Monotony, lack of power and responsibility	20	12	8
Pace of work, machine problems	—	3	11
Working and physical conditions	30	18	8
Other[a]	31	32	13[a]
Total	99	100	100

[a] Five percent concerned with low pay and lack of advancement.

about working conditions and the physical environment. Monotony, low job control, fast pace of work, and machine troubles, the complaints traditionally associated with low-skilled jobs (Goldthorpe *et al.*, 1968:21), together constituted less than a fifth of all complaints.

Only in OLDS were job complaints related to skill. As expected, the less skilled objected most to the pace of work and working conditions, but not to monotony and lack of job control. The machine operators, rather than assemblers, were most bothered by monotony. At PAL, machine operators and those in test-inspection-repair complained most about work pace. In short, indirect questions about job dissatisfaction slightly increased the complaints about monotony and lack of job control, but other complaints, such as working conditions, were still mentioned more frequently. Importantly, the complaints of assemblers closely resembled those of other workers.

Finally, the problem of monotony was approached directly by asking workers whether they wanted, without change in pay, periodic changes in type of work activity. Except for IKA, three-quarters or more wanted either no change in job routines or a change less frequently than once a month (see Table 6.11). In IKA, 43 percent desired more frequent changes,[7] a surprising finding in view of the fact that management changed job assignments of two-fifths of the workers once a month, a much higher ratio than in the other plants. PAL's employees most opposed change in job routines; seven-tenths wanted change less often

[7] Assemblers, 40 percent; machine operators, 30 percent; TIR, 39 percent; craftsmen, 45 percent.

TABLE 6.11 DESIRED CHANGE IN TYPE OF WORK ACTIVITY AND FREQUENCY OF CHANGE (PERCENTS)

Work Activity Change	*OLDS*	*FIAT*	*IKA*	*PAL*
No change desired	54	52	26	59
Change desired:				
More than monthly	17	11	26	—
Monthly	8	9	17	12[a]
Less than monthly	21	28	31	28
Total	100	100	100	99

[a] Monthly or more frequently.

than once a month. Since the pattern for OLDS and FIAT approached PAL's, IKA represented a deviant case. It may be that workers simply want more of the same; those who do not experience change want stability; those who change, want more of it.

Except for FIAT, the lower their skill and job control, the more workers wanted change in type of work routines. But variations around this pattern were found in each factory. In FIAT the percentage wanting change was almost the same for operators, assemblers, and inspectors, but only in OLDS did assemblers greatly exceed other workers in their desire for change. The extremely rapid pace of the line undoubtedly accounts for this situation. While the semiskilled in OLDS wanted more frequent change, in PAL more of the skilled wanted change. In PAL, skilled workers often turned out identical pieces similar to those produced by semiautomatic machines in OLDS, which were operated by semiskilled workers. PAL's machines were less automatic and more operations were required to make the same parts. In IKA, unskilled assemblers who worked on different lines had more variation in their work than semiskilled machine operators.

The findings on monotony are clear. Only a small minority of workers voluntarily mentioned it when asked about their job satisfaction. Less than a fifth mentioned monotony when asked what they specifically disliked about their jobs. More than half wanted no change in job routines, and those who did preferred infrequent changes. The desire for change seemed more related to how management utilized the workers' skills than to the routineness of work operations. Thus, assembly workers were not necessarily more concerned about monotony than others.

Social Interaction at Work and Job Satisfaction

In theory, the social environment of the work situation affects the job satisfaction of the skilled and the unskilled differently. Because of the freedom of movement the job provides, the skilled can make social contacts with other workers easily. Thus socially satisfying relations supplement the satisfactions derived from interesting work. For the unskilled, the monotony of the work means that job satisfaction must be derived mainly from contacts with work-mates. Yet social interaction may have little or no effect on the job satisfaction of the skilled, but some effect on the unskilled. To explore this idea I compared respondents with high and low satisfaction who experienced little or a great deal of interaction on the job.

In Chapter 5, I reported that the skilled and those with most control over their machines had the most interaction with their work-mates on and off the job. Yet job satisfaction was not related to the amount or quality of contacts with co-workers inside or outside the factory. To account for this apparent anomaly, I examined the relationship between skill level and job satisfaction, controlling for interaction. Except for PAL, no relationship was found for the skilled which suggests that interaction has little effect on job satisfaction when the work is interesting (see Table 6.12). But for the unskilled and the assemblers (except for PAL), more of the satisfied worked in high-density departments, had more interaction on the job, and more good friends among their work-mates. Machine operators and semiskilled workers revealed an opposite pattern: more of the satisfied worked in low-density departments, had little work interaction, but more contacts with work-mates outside the plant. Finally, workers in test-inspection-repair departments resembled assemblers; more of the satisfied had more work contact and more good friends among their work-mates. This pattern was strongest for PAL and weakened until it was absent at OLDS.

These findings support a sociotechnical explanation of job satisfaction (Blauner, 1964; Meissner, 1969). Since craftsmen find their work satisfying they do not need rewards from social interaction. The unskilled, on the other hand, find their work more satisfying when they have more opportunity to interact. Machine operators, who work in noisy, low-density departments, have little opportunity to interact on the job, but their satisfaction is increased by contacts with work-mates outside the factory. Em-

TABLE 6.12 JOB SATISFACTION ACCORDING TO AMOUNT OF WORK COMMUNICATION BY SKILL LEVEL (PERCENTS)

		Low Communication		*High Communication*	
Skill Level		*Not Satisfied*	*Satisfied*	*Not Satisfied*	*Satisfied*
OLDS:	Unskilled	33	67	18	82
	Semiskilled	12	88	19	81
	Skilled	8	92	7	73
	Total	17	83	14	86
FIAT:	Unskilled	35	65	14	86
	Semiskilled	26	74	24	76
	Skilled	8	92	7	93
	Total	28	72	16	84
IKA:	Unskilled	15	85	10	90
	Semiskilled	18	82	0	100
	Skilled	4	96	8	92
	Total	14	86	6	94
PAL:	Unskilled	42	58	42	58
	Semiskilled	27	73	26	74
	Skilled	33	67	13	87
	Total	34	66	27	73

ployees in TIR who work in crowded but quieter quarters can communicate easily, a condition which increases their job satisfaction. Such an environment existed in PAL but not in OLDS. There the mechanization of TIR operations had the effect of dispersing workers widely, reducing their interaction. The major exceptions to this sociotechnical interpretation were the assemblers and machine operators at PAL whose job satisfaction was little affected by the opportunity for interaction on the job. We have seen that Indian workers placed less importance on work satisfaction than did other employees and that PAL's departments were about equal in density. These two conditions may have had the effect of reducing interaction between the job satisfaction and social interaction variables.

CONCLUSIONS

The results of this study cast serious doubt on the ideology of the machine-haters. Even in the American automobile industry, where technology is allegedly most dehumanizing, workers expected satisfaction in their work. The situation differed little in the less industrial countries; autoworkers preferred working to leisure, not out of a sense of duty or a need for sociability, but because they thought that work ordered their lives. They did not see the factory as a restrictive and unattractive environment; on the contrary, most preferred its noise and smell to the antiseptic atmosphere of the office. Even rural migrants were not nostalgic about the farm and the urban-born rarely mused about the joys of farm life.

Autoworkers everywhere seemed unaware of the ideology that they should hate their jobs. Most found their jobs quite endurable. The skilled liked them for their interest, and the less skilled for both interest and money. Whatever their reasons, autoworkers were not seeking to escape the factory. They knew that other occupations paid better and had more prestige, but this knowledge did not make them dissatisfied. Though most of them hoped and many expected that their children would find white-collar jobs, these hopes and expectations did not result in dissatisfaction with their own jobs. Workers envisioned a different pattern of mobility for each generation.

Social life was remarkably alike in all four factories. The technological environment associated with automobile manufacturing is more varied than most people realize. Surprisingly, assemblers complained little about job monotony—no more than other workers—and the desire for change in work routines was quite limited. Although skilled workers had opportunity to talk on the job, sociability contributed little to job satisfaction. The opportunity for the unskilled to talk and move about on the job made it more satisfying, but socializing with work-mates outside the factory also contributed to satisfaction.

Both where industry is recent and where it is not, autoworkers responded quite positively to factory life. Contrary to some thinking, increasing industrialization does not make employees more antagonistic toward work, the factory, and industrial jobs. Routinized work does not appear to affect job satisfaction in simple and direct ways. While this study tends to support Blauner (1964), Shepard (1971), Fullan (1970), and others in their view that the less skilled are more dissatisfied with their jobs than the skilled, the important finding is that the differences among the skill levels vary enormously from one factory to another in the same industry. The response to technology is a much more complicated business than American scholars have believed. Person-machine relationships are affected by many external variables, such as the level of national industrialization, how management uses skills, how workers integrate their social life in the department, and other factors. Worker response to automated production, on which we have no data in this study, is perhaps even more complex than their response to mechanical production. As Seeman (1971) has suggested, the speculations concerning human responses to machines, from Marx to Marcuse, need thorough reexamination.

7

Autoworkers and Their Unions

ARTICULATING WORK AND UNION SYSTEMS

Social relations in the factory, while shaped by technology, do not carry over unchanged into the community. Technology's influence on workers' relations to groups outside the factory is typically indirect. A possible exception is the labor union, which is supposed to be interested in the work situation and how technology affects it. But the actual involvement of unions in the work situation varies by both industry and society. Some unions are continuously and deeply involved in shop problems and show little interest in external and political affairs. Other unions, deeply involved in national politics, almost ignore problems on the factory floor. However, automobile unions in most countries claim that they are interested both in how technology affects their members, and also in how politics affects them. It is not unreasonable, thus, to expect workers to evaluate union performance in these two areas. This chapter proposes a theory which explains how workers who are exposed to different technologies develop different expectations of union performance. It then considers the problem of whether these expectations vary with the problems unions face in societies which differ in degree of industrialization. Finally, the organization, ideology, and problems of each union are described to provide a background for interpreting the findings.

Scholars agree that labor unions of some form invariably accompany the appearance of large factories because conflicting strata (management and workers) have different interests and goals. To maximize their profits and plant efficiency, managers want obedient employees; workers want to restrict managerial power and to increase their own share of the profits. The organization of industrial production puts many workers in contact with one another and creates preconditions for unionization (Marx and Engels, 1959:16; Durkheim, 1964:23-31). Without the union, workers are impotent in an organizational society because only the union is exclusively dedicated to serving their interests in the factory and to giving them status and influence in society. Therefore, whatever the local technology, once unions appear, workers remain loyal members almost irrespective of the union's policies, ideologies, and effectiveness. I expected evidence of this loyalty in all four factories of this study.

Industrial Structure and Union Behavior

The impact of technology on union structure and worker behavior is widely recognized but little studied. Kerr and Siegel (1954) early observed that unions in the same industries (e.g., mining and shipping) behave similarly (e.g., are equally militant) in different societies because they respond to the same technological and organizational problems. All unions have three main functions: grievance bargaining with supervisors to resolve problems in the factory; collective bargaining with management to settle issues over wages, hours, and working conditions; and external bargaining with other interest groups in the nation for political influence. The union's goal is to extract as much as possible from supervisors, management, and government.

Union behavior in the factory is conditioned by the technology of the industry. To illustrate this, two industries are examined: an industry where most workers are skilled (e.g., machine tool) and a mass production industry (e.g., automobile) where most are semiskilled and unskilled. In the machine tool industry, skilled machinists have considerable control even without their union because their work cannot easily be rationalized or replaced (Lipset *et al.*, 1956; Stinchcombe, 1961; Sayles, 1963). Common background, common training, and the ability to form solidary work groups enable craftsmen to arrive at consensus on how to handle work-related grievances. Since they need little help to handle their immediate supervisors, union officers are expected to concentrate on collective bargaining with management; e.g., wages, technological change, and related issues. Skilled workers think they know all the issues and all the answers. They are quick to evaluate union officers. Under special circumstances, craftsmen take into account the political effectiveness of the union, but generally they evaluate performance in terms of success in dealing with management (Aronowitz, 1973: 184-194).

Almost at the opposite technological extreme are unions in

large mass-production industries. Because most workers have few skills and are easily replaceable, managerial power would be almost unlimited unless checked by strong unions. Often the employees of these large factories do not have a common background of culture, training, and experience. Their wide dispersion in the factory and community reduces their ability to communicate quickly and arrive at consensus. Moreover, the technology and ecology of the departments (e.g., assembly, presses, automatic machines) restrict their interaction and ability to form solidary work-groups. In short, compared to the skilled, mass-production workers exhibit low social solidarity—which, in turn, makes them more dependent on their officers. Power becomes centralized in the hands of officers who handle problems at all three levels: local grievances, collective bargaining issues, and political problems.

Mass-production industries with high fixed costs increase their profits by increasing labor productivity. Union officers monitor supervisors to make sure that production quotas are not exceeded. Grievances (e.g., speed-up of the line) are not resolved by on-the-spot consultation with the foreman because grievances are often categoric and not individual matters. An elaborate grievance machinery involving increasingly high echelons of the union and management handles thousands of complaints annually. Since skill differences tend to be small, collective bargaining can proceed efficiently on a factory-wide or industry-wide basis. Productivity, unemployment, overtime, and pace of work are common collective-bargaining issues. Negotiations are often drawn out, and strikes, when they occur, are tests of endurance. Officers try to supplement collective-bargaining gains with government-assisted benefits in unemployment, health, and old-age insurance, as well as subsidized housing, price controls, and lower taxes. To be politically effective, officers of mass production unions need militant, politicized, class-conscious workers.

Technology, Grievances, and Collective Bargaining

All labor bodies in the four factories were the industrial type of unions found in mass-production industries. Pressure for increased labor productivity was highest in the factories which introduced most technological changes, OLDS and FIAT. Therefore, one would expect OLDS and FIAT workers to be most concerned with the union's effectiveness in handling shop grievances over working conditions, the pace of work, production quotas, breakdown of machinery, and supervisory authority. Because technology was stable, workers in IKA and PAL should be more concerned with the union's non-shop functions: collective bargaining for wages and fringe benefits, and perhaps political bargaining for government-financed benefits.

Since assemblers and machine operators were more exposed to production speed-ups and technological changes than were inspectors and skilled workers, the former should press unions to improve working conditions. Almost two-thirds of the labor forces of FIAT, IKA, and PAL were assemblers and machine operators; OLDS had somewhat fewer. These inter-plant variations were not sufficiently large to produce different amounts of worker pressure on union officials to pursue different goals. Restricted mobility and communication may also stimulate grievances over working conditions. Data in Chapter 5 show that assemblers and machine operators in FIAT and IKA were least able to move about and talk. PAL employees were the most mobile, followed by OLDS's, but in both factories they were able to talk to about the same number of people. On the basis of technological pressures, OLDS and FIAT workers should be most concerned about working conditions. On the basis of mobility and communication restrictions, FIAT and IKA workers should be the most concerned. Since FIAT's employees were exposed both to sophisticated technology and to restricted mobility, they should be expected to press their unions hardest to deal with working conditions. OLDS, IKA, and PAL should follow in order.

I see no reason why the technological differences among the factories should cause workers to put different amounts of pressure on union officers to wrest economic concessions from management. Since low living standards and primitive technology go together, workers in the less industrialized countries might exert more pressure on unions to bargain for economic benefits; PAL's workers would press most, followed by FIAT, IKA, and OLDS. I see no theoretical basis for arguing that variations in plant technology affect the pressure which workers put on unions to seek economic benefits through political action. It is true that unions in the most industrialized countries have had the most time to bargain and create institutions to meet economic and other crises (Smelser, 1963a). In less industrialized countries the briefer experience of unions with collective bargaining may lead them to exert more pressure on government to deal with economic problems. If this is the case, the Indians should exert most pressure on their unions to seek political solutions, followed by Argentinians, Italians, and the Americans (Millen, 1963).

Finally, within each factory, workers who perform different operations may press their union to pursue different goals. The unskilled are least affected by technological change. They evaluate their jobs in terms of the quality of supervision and press the union to protect them from arbitrary authority. Semiskilled workers are most exposed and least resistant to technological change. They are most concerned about the pace of work, the condition of their machines, the piece rate, and matters affecting working conditions. They press unions to monitor the employer's adherence to the contract and to improve working conditions. The solidarity of the skilled, on the other hand, enables them to deal directly with supervisors over working conditions. Therefore, they call on the union to consolidate and maintain the gains they have already achieved. In short, although all workers urge unions to bargain for economic benefits, the unskilled press more for supervisory control, the semiskilled for technological control, and the skilled for the preservation of privilege. The more industrialized the society and the more complex the technology, the more these trends should hold. How workers respond to the political functions of unions merits separate treatment.

Technology and Union Politics

Political unionism is the belief that the primary function of the union is to change the political and social system of the society for the benefit of the working classes, the belief that main improvements are brought about through political action rather than by bargaining with management. The less industrial the society, the more this seems to be the case. In the industrialized world, American unions represent a deviant case because they attach relatively little importance to political unionism (Millen, 1963). Whatever the formal political ideology of unions, the higher that officers are in the hierarchy, the more they are dedicated to political unionism (Wilensky, 1956:244).

The failure of union leaders to activate workers politically has been the despair of liberals and radicals since Marx's time, for if workers could be mobilized, unions might be able to bring about massive social changes. Many reasons have been adduced for the failure of workers to heed political advice of their officers: socialization in bourgeois institutions, false class consciousness, authoritarian personalities, and repressive legislation. These explanations ignore the realities which workers everywhere face on the job.

The disagreement between officers and workers on the priority of union functions is a normal consequence of industrial and union growth. Officers realize that survival in an interdependent economy depends upon the union's ability to control the forces which affect employment. Unions must cooperate with other unions and with political parties which share their interests in the political economy. As in all organizations, the higher the officers are in the hierarchy, the more they focus on external and political problems and the less they know about local concerns. To be effective politically, union officers must convince workers to contribute money to the party and to vote for it. Officers must convince the rank and file that union political influence will result in economic rewards.

Why should workers find this difficult to understand? Since they spend more time on the job than in other activities, they are more concerned with the job than anything else. By virtue of their socialization, workers live in an organizational desert (Spinrad, 1960); they do not learn how the political system works and how to change it (Litt, 1963:74). In political conflict, the upper strata have the advantages. They not only know how the political system works but they know how to manipulate it to their advantage. Moreover, the upper strata need fight only defensive battles to maintain the status quo (Gamson, 1966:122). This requires less time, energy, and money than does changing institutions. In short, for union officers to expect workers to have a sophisticated view of the political process is unreasonable because workers see little connection between the union's activities in the factory and the political activities of high union officials (Goldthorpe *et al.*, 1968:31). While workers may approve of and support friendly political parties, they hate to see the union spend their dues in a political game which has uncertain benefits at best.

When industrialization begins, when workers move from subsistence to cash markets, they want economic security and protection from arbitrary management rule (Belshaw, 1965:115-122;

Moore, 1963:347-451). The weakness and inexperience of unions at this stage lead workers to recognize the importance of social solidarity (Smelser, 1959:313-322). In mature industrial societies, where unions are accepted and strong, wage increases become almost routine (Galbraith, 1967:247-281). But other pressures mount. Increasing work rationalization, improvements in technology, and rising production quotas shift the worker's attention to improving working conditions (Blauner, 1964:98-106).

Unskilled workers in industrial societies resemble workers in newly industrializing countries. The latter are poorly educated, economically insecure, and organizationally unsophisticated. Apathetic toward unions and politics, they try to avoid conflict in order to stabilize their incomes (Chaplin, 1967:146-167). However, they are easily pressured and politically unreliable because they are capable of simultaneously embracing both radical and conservative political ideologies (Morris, 1968:512-514). Urban-born semiskilled workers are occupationally immobile, but better educated and politically informed. Their response is to become the most militant unionists and the most radical members of the working class (Lipset and Bendix, 1959:64-72). Skilled workers, in contrast, control most of the conditions which affect their welfare (Blauner, 1964; Meissner, 1969). They are often recruited from privileged backgrounds, become the most influential segment of the working class, and control the local union for their own advantage (Michels, 1959:292-294; Giddens, 1973:204). Since they usually get what they want through peaceful bargaining, they prefer to avoid militant tactics. As the most involved section of the working class in plant, union, and community affairs, they frame the issues of their organizations, dominate the flow of communication, and exercise a self-interested and often conservative political influence (Lenin, 1943:92).

In summary, regardless of the industrial development of a society, workers support their unions because they have no choice. The technology and organization of automobile manufacturing tends to throw the power to handle grievances, collective bargaining, and political action in the officers' hands. Differences in technologies of the factories under study suggest that OLDS and FIAT workers should be most concerned with working conditions, and the others with economic rewards. Though all workers probably eschew political unionism, semiskilled workers, especially, in the less industrial countries are most attracted to it. Skilled workers dominate the informal communication system in the factory and exert a self-interested and often conservative influence on union affairs. My effort to specify how technology affects worker preferences for different union functions is offered with a good many reservations. As Hamilton (1965) pointed out, abundant literature can be found to support almost any position.

Union Structures and Ideologies

The unions in the four factories differed in their histories, types of bargaining structure, autonomy from national unions, grievance procedures, internal factionalism, dues structure, power vis-à-vis local management, party ties, and political ideology. Some of the selected attributes of the unions are summarized in Table 7.1 and described below. If, despite such variation in union structures, workers in different societies behave toward their unions in accord with the above theoretical speculations, we may have considerable confidence in the development hypothesis.

Table 7.1 Some Important Characteristics of Labor Unions in Four Automobile Plants

Characteristics	*OLDS*	*FIAT*	*IKA*	*PAL*
Name of union	UAW	CGIL, CISL, UIL, SIDA	SMATA	EMS, INTUC
Economic bargaining	National	National and local	Local	Local
Grievance handling	Rationalized	Semi-rationalized	Rationalized	Personalistic
Dues structure	Check-off	Voluntary	Check-off	Normative
Power vis-à-vis management	High	Moderate	Moderate	Low
Wage structure rationality	High	Medium	Medium	Low
Political ideology	Liberal	Conservative to radical	Laboristic	Socialistic
Ideological insistence	Low	Very high	Moderate	High
Party linkages	Moderate	Strong	Moderate	Moderate
Factions	Weak	Strong	Moderate	Weak
Labor force	12,000	33,000	11,000	8,200

OLDS

About 9,000 employees of Oldsmobile division of General Motors constituted a separate local of the United Automobile Workers of America (UAW). The UAW was a financially strong, aggressive union which had been more successful than average in obtaining concessions from management. Compared to locals in Detroit and Flint, the Lansing local's relations with OLDS management had been calm for over a decade. A national contract governing all General Motors divisions was periodically negotiated with the International Union. It specified wage rates for every job, lay-off rules, grievance procedures, seniority rights, shift allocation, and other matters. Local issues, such as the pace of the line and working conditions, were left for OLDS management and Local 652 to negotiate. A full-time representative of the International assisted the local and maintained communication with headquarters. Legally, union membership is voluntary; in practice, almost all employees were members and management sent workers' dues directly to the union.

From the late 1930s to about 1950, two rival factions were active in the Local; the Reuther caucus, named after the then-president of the International, and the anti-Reuther caucus. The latter comprised political radicals who favored local autonomy. When Reuther's forces gained full control of the International, local caucuses disappeared. Unlike those in the International, officers in the OLDS local typically were unseated after one or two terms in office. Interest in local elections was rarely spirited, as evidenced by the fact that less than one-quarter of the members voted in them.

The UAW was the largest union in Lansing. Its leadership traditionally supported the Democratic Party. Although the law forbade the union to spend dues on political activities, its Committee on Political Education solicited campaign funds and actively worked for Democratic candidates. In these activities it did not have the solid backing of the membership. About three-fifths did not go along with the political recommendations of the union, an equal percentage felt the union should not endorse political candidates, and four-fifths thought that union dues should not be spent for political action. Finally, two-fifths identified themselves as independents, Republicans, or political conservatives. The UAW, along with other locals, backed selected non-partisan candidates for city offices and the Board of Education. Although its record for supporting successful candidates was good, the unions were never able to elect one of their own members (Form, 1959).

The UAW traditionally encouraged its locals to place representatives on the boards of all municipal and private welfare agencies. Although labor achieved widespread representation, it never gained a board majority and rarely did labor elect one of its members to chair a board. A study of labor and community (business) influentials in Lansing revealed that the two groups differed little on most local issues. Labor appeared more concerned with board representation than with developing a distinctive program to help the poor and ethnic minorities (Form and Sauer, 1963).

FIAT

Four industrial unions belonging to different labor confederations were active in FIAT. Like most European unions, each was unofficially attached to a political party and dedicated to a distinct ideology and political program. Union membership was not compulsory; in FIAT only 15 percent of the workers were thought to be dues-paying members. Yet local elections were "compulsory," and about 90 percent of FIAT's employees voted annually for slates of local candidates who represented the four unions. In this situation, the worker could avoid an ideological commitment only by casting a blank ballot. Winners of the election, on a proportional-representation basis, constituted La Commissione Interna, a local body which interpreted the national contract and settled local grievances (Neufeld, 1954; LaPalombara, 1957). The Commissione is supposed to represent all workers and confront management as a body, but this seldom happened in FIAT. Management typically conferred with each union separately and refused to meet with the CGIL "communist" delegates. Since the unions typically were unable to agree on anything, management was in the enviable position of pitting the unions against one another and getting the credit for distributing benefits.

The Italian labor scene is complex and confusing. A brief description of each union may facilitate interpretation of subsequent data. The oldest, strongest, best organized, and most inclusive labor federation is the Confederazione Italiana del Lavoro (CGIL) which is allegedly dominated by communists and left-wing socialists. Automobile workers belong to FIOM (Federazione Impiegati e Operai Metalmeccanici), the metalworkers' union. FIOM–CGIL was the largest union in FIAT, and it polled 30 percent of the votes in the last election. Over the years, CGIL had lost representation in the Commissione because management and competing unions accused it of being willing to sacrifice worker well-being for Communist Party goals. FIOM denied this accusation. It was easily the most aggressive union, calling for strikes to end management's labor-splitting tactics, its policy of hiring skilled workers from outside rather than promoting them from within, its paternalistic policy of providing services (e.g., medical, housing) rather than wage increases, and its tying the "premium of collaboration" to production rather than profits.

The second largest union in FIAT was SIDA (Sindicato Italiano dell' Automobile) which received 28 percent of the votes. Not tied to a political party, it was exclusively a FIAT union. Originally, its leaders were members of Catholic CISL, but they opposed its socialist, militant, political stance. To counter the charge that it was a company union, SIDA joined the European International Catholic Union Confederation headquartered in Belgium. SIDA officials claimed that they represented the Catholic European democratic movement, which was not linked to any party. They insisted that FIAT's employees needed a union of automobile workers separate from the larger metalworkers' union, to attack the unique problems of the industry. The most important local problems SIDA leaders mentioned were: seasonal changes in production, technological changes in the industry, the adjustment of new migrants in the factory, and working conditions in various FIAT departments.

The third largest union at FIAT was UIL (Unione Italiana del Lavoro) which polled 23 percent of the votes. Originally, UIL was part of the short-lived FIL (Federazione Italiano del Lavoro), an anti-communist non-Catholic social democratic union which broke apart in 1950. UIL's program is admittedly similar to that of Catholic CISL, but UIL refused to be associated with the Church. The main issues UIL leaders saw at FIAT were: reduction of daily and weekly hours of work, the need for a third week of vacation with pay, and the reorganization of the job classification, collaboration, incentive, and productivity schemes. UIL, like SIDA, wanted to confront the unique problems of the automobile industry and pay less attention to class politics. As a democratic socialist union, UIL opposed CGIL's subservience to the Communist Party and CISL's allegiance to the Church.

CISL (Confederazione Italiana Sindicati Lavoratori), the Catholic labor federation, was the second-largest labor confederation in the nation. In an effort to end economic chaos, it split from CGIL soon after World War II to support the Christian Democratic Party. The metalworkers' branch of CISL was FIM (Federazione Italiana Metalmeccanici), which had the support of many Catholics, republicans, Christian socialists, and some social democrats. In 1953, as FIOM weakened in FIAT, FIM, with management's help, won the majority on the Commissione. FIM's leaders became convinced that management was firing communist employees as part of a plan to weaken all unions and to institute a company union. The more aggressively FIM acted, the more strikes it threatened, and the more it united dissident groups, the more management shifted its support to other unions. FIM's strength declined and, at the time of the study, it was the smallest union, garnering only 16 percent of the votes in the last election. FIM's socialist-oriented leaders were almost as militant as FIOM's. Both unions felt that SIDA and UIL were playing into management's hands by accepting small gains, by refusing to strike, and by not uniting all unions to fight for fundamental reforms.

As is often the case where factionalism runs rampant, leaders of the various unions agreed on many issues but did not know it. They all wanted the company to hire additional employees because workers were tired of the five-and-a-half-day week and the excessive overtime. All unions agreed that job classifications were being violated: some skilled workers were doing semiskilled jobs but more semiskilled were doing skilled work. Management clearly profited from legal arrangements which froze workers in skill levels. All leaders agreed that FIAT made enormous profits and could pay higher wages. Most felt that FIAT's social services were paternalistic, and that the company should pay higher wages instead of providing services. Last, officials wanted the "premium of collaboration" (a bonus linked to productivity increases) to be guaranteed.

Although only 12 percent of the workers in the sample admitted being union members, union sympathy was pervasive. I wanted to know which union and party individual workers identified with. But FIAT's extreme sensitivity on union issues forced me to approach the subject with great caution (see Appendix B). I carefully arranged interview questions in a way that would encourage workers voluntarily to reveal their union preferences, but I did not succeed. Midway through the interviewing, I asked interviewers to pose the question directly, and most workers answered it. Unfortunately, union identification was obtained for only 44 percent of the respondents.

A comparison of the union identification of workers in the sample and the official voting records in the departments showed a severe underrepresentation of radical CGIL sympathizers and an overrepresentation of the Catholic CISL. The data in Table 7.2 reflect management's policy to concentrate left-wing sympathizers in selected departments (Ornati, 1963:524) and possibly the reluctance or fear of workers to reveal their CGIL sympathies. In the interviews, some CGIL sympathizers undoubtedly identified with the militant Catholic CISL, a respectable thing to do because the union was associated with the dominant Christian Democratic Party which even FIAT backed in national politics. Finally, up to 20 percent of the workers in one part of the interview revealed hostility or apathy toward unions. When pressed for a union preference many undoubtedly opted for the "safe" unions: SIDA or UIL.

TABLE 7.2 UNION IDENTIFICATION FOR FIAT, THE SAMPLED DEPARTMENTS, AND THE SAMPLE

Unions	*Plant*	*Departments Sampled*	*Workers Sampled*
CGIL—Left	30	20	10
CISL—Catholic	16	14	26
UIL—Free	23	26	29
SIDA—Independent	28	40	35
Total	97[a]	100	100

[a] The remainder voted for CISNAL, a Fascist union.

The interviewers reported that most workers were interested in unionism, sympathetic to one union, impatient with multi-unions, and unwilling to support unions financially. Yet all four unions employed full-time officials and secretarial staffs. They all had relatively large and well-equipped offices, and published newspapers. "Informed observers" reported that the Communist

Party supported CGIL, the Church supported CISL, the Americans supported UIL, and FIAT supported SIDA. Whatever the source of funds, the bulk of union work was done by full-time staffs. Membership meetings were rarely held and very poorly attended. Less than 5 percent of the workers ever participated in any union activity (e.g., sports, rallies, political discussion groups, picketing).

A few conclusions may be drawn about the union situation at FIAT. In terms of traditional radical-conservative schemes, the unions line up as follows: CGIL (communist), CISL (Catholic), UIL (social democratic), and SIDA (independent). The flow of union power at FIAT shifted over the years from CGIL to CISL to SIDA, with UIL showing strong recent gains. All unions agreed on certain issues, but disagreed on their relative importance and how to deal with them. Union competition gave management inordinate power to influence the union sympathies of workers and to stimulate union rivalries. Finally, the changing patterns of union voting over the years reflected a confusion or fragility about ideological matters among many workers.

IKA

The labor situation in Argentina cannot be understood outside the context of Peronism. Whatever their feelings about Peron, scholars agree that he was responsible for integrating Argentine workers into the national political scene (Fillol, 1961:55). Peron reorganized the Confederación General de Trabajo and instituted a series of benefits: holidays with pay, a forty-hour week, minimum wages, social security, retirement pensions, and medical benefits (Pendle, 1963:123-126). After Peron's ouster, labor-union influence in national politics was greatly reduced. Post-Peron governments regarded unions as semi-legitimate, even illegitimate, sources of political opposition which had to be heavily monitored (Davis and Goodman, 1972:167). Governments paid little attention to public housing, unemployment benefits, and other welfare services, with the consequence that the relative economic position of the working class deteriorated. Though workers and union officials disagreed on whether Peron should return or what Peronism really meant, most agreed that they were Peronistas in the sense that they wanted the unions to be part of the legitimate "political combination" to improve the economic position of the working class. No important union leader could afford to be anti-Peronista. Although socialist and communist sympathizers were active in the labor movement, the central figures were Peronistas. For most workers, to be actively pro-union and Peronista was to be anti-government and pro-working class. In short, Peronism represented for most workers a laboristic, if not a radical political, ideology.

The Union in IKA's Córdoba plants was SMATA (Sindicato de Mecanicos Automotores y Trabajadores Afines), an industrial type of national union with headquarters in Buenos Aires. However, the Córdoba local was its largest and strongest affiliate, and it operated rather independently. Employees in the other automobile plants in Córdoba (FIAT and DINFIA) did not belong to SMATA. Their unions were regarded as weak, possibly because they did not join forces with the strong Córdoba *sección* of SMATA. Although IKA was not a closed shop, about 85 percent of its manual workers were members of the union and almost all of them voted in union elections. One percent of the workers' wages was taken from their pay by management and turned over directly to the union. The union owned its own headquarters and was building a 2,000-seat auditorium.

Compared to other Argentine unions SMATA had been very active in the factory, somewhat along the pattern of American unions. National officers in Buenos Aires however, were typically and heavily involved in Peronist politics (see Alexander, 1962). In an attempt to integrate members into the union, SMATA provided small birth, death, and marriage premiums, discount arrangements at local stores, a barbershop, newspaper, pharmacy, and a recreational program. It also conducted leadership training courses. However, only a tenth of the workers participated in the union's sports program and most did not even know the location of union headquarters.

The main work of the local union was done by the Comision Directivo (the general committee) and three other committees. The Committee to Interpret and Apply the Contract met with management twice yearly. Its primary task was job classification; machines were rated for the complexity of skills required to operate them and the pay operators should receive. The Parity Committee met with management three times a year to determine how wages should be adjusted to the cost-of-living index. Although the index was set by a government bureau, the committee sought increases above the index as well as other benefits. The internal Grievance Committee met twice weekly with management to settle problems of pay, safety, penalties for absenteeism, supervisory authority, work rotation, and adequacy of services (health clinic, cafeteria, etc.). In addition to the committees, 265 sub-departmental delegates were elected for two-year terms to represent workers on the plant floor.

Three factions in the union were based on political sympathies. A socialist-communist faction criticized the unaggressive "pro-management" stance of the current officers. It pointed to the low percentage of hours lost by strikes over the past few years (0.3 percent) as evidence of management's domination over the union. The Peronista faction wanted the union to become more active in politics and to involve government in labor disputes. Leaders of the moderate faction in power felt that continuous non-conflictful bargaining had raised wages to the highest level in the industry and community. They saw no need to involve third parties in labor disputes nor to risk long strikes which management could survive better than the workers could. Management supported this "non-political," plant-centered union, a fact not ignored by the other factions. Toward the end of the study the moderate faction lost power but the new leadership had not fixed a clear policy course.

Leaders of all factions agreed on some issues the union should confront: job upgrading and promotion, leveling production to eliminate periods of unemployment, increasing welfare benefits, and gaining wage increases above the cost of living.

PAL

During the Indian struggle for independence, the trade-union movement was primarily involved in fighting colonial rule. Communist, socialist, and Congress party labor leaders concentrated on a common cause until independence was achieved. Soon after, three trade union organizations were formed along party lines (Ornati, 1955:120-129). The old AITUC (All India Trade Union Congress) remained firmly in the hands of the communists; the Hind Mazdoor Sabha (HMS) became the refuge of the socialists; and the Gandhian INTUC (Indian National Trade Union Congress) clung to the ruling Congress party. About a third of the industrial labor force of the country joined these unions. All of them are highly centralized and dominated by their officers, who appear to have permanent tenure (Myers, 1959:34-37; Sheth,

1968:163-167). AITUC calls for violent, militant class-conscious unions to change India into a communist state. INTUC goes along with the gradualist nonviolent evolutionary socialism endorsed by the Congress Party. HMS emphasizes the necessity of building class-consciousness unionism through strikes and passive resistance, with the ultimate aim of seizing control of the government and industry through legitimate democratic means.

According to Indian law, more than one union may represent workers in a plant. The law does not specify how the bargaining unit is to be determined, nor does it require employers to bargain with the unions. Collective bargaining is extremely limited and unions rely on government tribunals, wage boards, and even legislation to settle wage and other disputes. Government "model standing orders" provide the framework for much collective bargaining (Bureau of Labor Statistics, 1961:29-30). Formal grievance procedures are rarely used. Typically, the worker goes to his foreman or the firm's labor or welfare officer to settle his grievance. Lacking satisfaction, he may turn to a union officer, who is not likely to be a co-worker acquainted with the local situation (Myers, 1958). Whatever channel he uses, disputes are settled on *ad hoc* grounds rather than by applying the provisions of the contract.

INTUC first held power at PAL, but it dwindled. In 1958 a strike of 110 days was begun by the Engineering unit (EMS) of the socialist HMS (Ministry of Labour and Employment, 1960). Although the strike was lost, management recognized HMS as the official representative of the workers. Accurate membership records were not available, but EMS probably represented about 75 percent of the workers, INTUC about 10 percent, and 15 percent were not members of either union.

Union membership required little involvement; elections were not held periodically and membership "meetings" were rare. Union officers in Bombay simply "nominated" active and loyal members in PAL to serve as local "leaders" and informed the workers of this fact. The leaders collected dues, passed out information on union policy, and gathered information on the activities of rival unions. Officials and leaders met once a month to discuss local problems. Open-air membership meetings were held occasionally on factory grounds, where officials would present their analysis of the situation, make a recommendation for action, and ask for membership support. Meetings were typically terminated without discussion.

We could not determine the strength of factions within EMS. Some workers who backed EMS objected to the absence of democratic procedures. About a third of the members of EMS and a few others volunteered the information in the interviews that the union and the company had joined hands to reward EMS leaders and other long-tenured loyal EMS members. Union leaders felt that the most important, long-standing issues at PAL were: (1) the chaotic job and wage classification scheme, (2) the unclear grounds for determining the bonus, (3) the poor transportation facilities to get workers to and from the factory, and (4) unsatisfactory leave, seniority, and promotional policies.

Discussion

The unions in the four plants can be ranked in terms of the extent to which the local officers were involved in both settling grievances and collective bargaining. Clearly, SMATA in IKA had most local autonomy, for its officers had major responsibility both for settling grievances and bargaining for wages. Local UAW officials in OLDS controlled grievance handling but economic bargaining was done in Detroit on an industry-wide basis. In FIAT, wages were basically set by the national contract covering all metal industries, but some supplementary bargaining took place locally. Grievances were handled by the Commissione Interna, which was not as strong as the grievance committees in OLDS and IKA. Finally, EMS officials in PAL were rather distant from the local scene in both the bargaining and grievance arenas. Wages were set by a mixed system of collective bargaining and *ad hoc* individual arrangements. While wages were adjusted to skill levels at entry, seniority and personal intervention by management and the union determined individual wages at later periods. Some grievances were settled by union intervention, but no written procedure for handling grievances had been established.

8

Technology, Unions, and Political Ideology

In the previous chapter a theory was suggested to explain the union beliefs and behavior of industrial workers in societies at different industrial levels. This chapter attempts to test that theory. I have argued that whatever the extent of manufacturing in a society and whatever their jobs, industrial workers typically support the union because it is their only organizational weapon. They believe that this weapon should be used to improve their wages and working conditions and not for political purposes. The less industry in a society, the more workers emphasize the economic functions of unions; the more complex and changing the technology, the more they want the union to improve working conditions. But everywhere, the unskilled are more concerned with economic gains and protection from arbitrary supervision, the semiskilled are more concerned with improving working conditions, and the skilled with preserving their economic and other advantages. Officers tend to place more stress on political functions of the union than do the rank and file, and the less industry in a society, the more this tends to be so. Skilled workers everywhere dominate the local union. Under special circumstances, they may exert a radical influence, but generally they are more conservative than other members. The skilled are more successful in shaping union goals because they are more active members and dominate political communication in the factory.

Union Involvement

The involvement of the member in the union is examined in terms of his evaluation of the organization and his participation in its activities. I have suggested that members approve of the union, whatever its politics or performance, because they know that only the union is exclusively devoted to their interests. Data in Table 8.1 show that this is the case; about nine-tenths of the employees in each factory felt that unions were necessary organizations and three-quarters were favorably inclined toward them. Yet a much lower percentage (less than half) believed that the local union was performing effectively and even a smaller percentage was highly interested in union affairs. Thus, though most workers believed unions to be a legitimate part of the industrial scene, interest and evaluation of the local union's performance varied considerably.

Data on the behavioral involvement of workers in unions are difficult to interpret because membership, voting, and participation in the union are structured differently in various countries.

TABLE 8.1 EVALUATION AND PARTICIPATION IN UNIONS (PERCENTS)

Items	*OLDS*	*FIAT*	*IKA*	*PAL*
Union Evaluation				
Unions seen as necessary	95	86	92	91
Favorable toward unionism	83	81	69	77
High evaluation of local union	45	25	66	50
High interest in unions	15	11	12	33
Union Participation				
Membership	100	12	91	88
Attended half or more of the meetings	7	5	32	37
Voted in last election	63	94	74	—[a]
Identified one or more union officers	51	39	94	95
Participated in union activities	22[b]	5	4	—[c]

[a] Not available, elections rarely held.
[b] Mostly in sports and recreation.
[c] No union activities.

Thus, membership was almost universal everywhere except in FIAT. However, the extent of voting varied widely; it was almost compulsory in FIAT, normative in IKA, voluntary in OLDS, and a rare ritual in PAL. Attendance at union meetings was low everywhere, except in IKA, where it was moderately high. Less than 10 percent attended half or more of the meetings in OLDS and FIAT. PAL's higher figure is misleading because meetings were not held regularly. On rare occasions during the lunch hour, union officials addressed the workers in the factory yards. Half of PAL's employees thought the union met once a year, less often, or never. Only 2 percent of those who attended ever spoke at a meeting compared to 25 percent for OLDS. Though more workers in IKA and PAL than in OLDS and FIAT were able to provide the names of officers, this was probably because officers were more visible to members in the smaller factories. Considering all the evidence, union involvement was probably highest in IKA followed by OLDS; FIAT and PAL were about the same. The general low degree of union involvement conforms to the findings of many American studies (Spinrad, 1960).

Apparently the pattern of involvement in the union gets set quickly upon industrialization. Workers quickly understand the need for unions, but the local union is evaluated in terms of its performance. Only in IKA did a majority of members judge performance to be good. Approval was slightly lower in PAL and OLDS, but in FIAT, a majority judged performance to be poor (see Table 8.2). In short, evaluation of the local's performance was not related to support of unionism in general. FIAT's employees were critical of both the local's performance and unionism in general, IKA's were approving of the local's performance but most disapproving of unions in general, and both OLDS and PAL workers judged their local's performance as lower than unionism in general.

The satisfied gave relatively few reasons why they thought their unions performed satisfactorily, but the dissatisfied gave many. Expectably, the main favorable reasons were union success in economic bargaining and good labor relations: that is, no strikes. Negative evaluations, especially in FIAT, focused on internal union problems such as poor leadership, factionalism, inefficiency in dealing with grievances and negotiations, and overemphasis on politics. These evaluations seem to reflect the strength of the unions. OLDS and IKA had relatively strong unions, while FIAT and PAL had divided and weak unions. Importantly, union ineffectiveness was blamed on internal organizational problems and not on management. In PAL, the union was criticized for its inefficiency, corruption, but especially for poor handling of grievances. In FIAT, criticisms centered primarily on union rivalries and inefficiency.

But FIAT employees were critical of the national union scene as well of the Commissione Interna. Though about three-quarters of the employees identified with one of the four unions, one-fifth were critical of the effectiveness of their union in the nation, in

TABLE 8.2 EVALUATION OF LOCAL UNION PERFORMANCE (PERCENTS)

Items	*OLDS*	*FIAT*	*IKA*	*PAL*
Evaluation				
Good and very good	45	25[a]	66	50
So-so	37	25[b]	30	32
Not good and poor	18	50	4	18
Total	100	100	100	100
Reasons				
Positive				
Good economic performance	15	8	19	36
Good labor-management relations	17	7	17	5
Other positive reasons	15	12	27	16
Negative				
Internal organizational problems	25	51[c]	11	22
Poor grievance and contract handling	7	6	6	19[d]
Union too political	—	7	14	—
Other negative reasons	21	9	6	2
Total	100	100	100	100

[a] Responses to the question, "What is your evaluation of the operation of the Internal Commission?" Evaluation of the union with which the worker identified was: favorable, 77 percent; mixed, 19 percent; and negative, 4 percent. Almost one-quarter of the workers were not identified with any union.

[b] "Assets and defects," but the reasons given here were overwhelmingly critical.

[c] About 17 percent responded "inefficient."

[d] All responded, "poor grievance handling."

Turin, and in the company. But they favored the local more than the national union; 70 percent approved of the local's performance in FIAT, in contrast to 55 percent for the national union. The objection on the local level was structural in the sense that workers blamed the Commissione Interna as a system more than they blamed their own unions. Only one-quarter considered the Commissione Interna to be an effective device to handle factory problems, one-quarter thought it had both assets and defects, and almost one-half evaluated it as basically defective. Most critics of the Commissione felt that it was designed to keep unions disunited while a smaller proportion felt that the Commissione was inherently an inefficient mechanism for problem-solving. But basic to both positions was the criticism that unions were too concerned with ideological problems rather than problems of the worker.

Priorities of Union Tasks

In interviews with union officers, we obtained information on their goal priorities for the union, their political ideologies, and related matters. Four officials from each union were selected, except at FIAT, where we interviewed twelve, three from each of the four unions. The highest national or regional officer in the city was interviewed; the remainder were local officers. In addition to providing information on the history of the union and important current issues, they were asked to rank four major union tasks: economic bargaining, improving working conditions, building social solidarity, and changing the political and social system of the nation. Almost all officers replied that all tasks were important. When pressed to rank them, only five of the 23 ranked economic bargaining as most important and all five were local officials. Then we asked the officers to rank the importance of economic and political goals; 17 of the 24 insisted that they were equal and could not be ranked. Only SIDA officials in FIAT thought that unions should not engage in political action. Of the others, only UAW officials explicitly rejected the idea that the goal of political action was to change the political and social system of the country. In short, except for SIDA and the UAW officers, all supported the ideal of political unionism (cf. Landsberger *et al.*, 1964). FIAT unions placed greatest stress on political action, followed by PAL, IKA, and OLDS.

Workers also were asked to rank the importance of union tasks or functions. They were asked: first, to name the most important union functions; second, to rank the four stated union functions presented by the interviewer; third, to name the most important local problems the union should face; and last, to rank the problems which their local officials thought were most important. The most frequent response to the first unstructured question in all four plants was, "defend the interests of the workers," and the next was "economic bargaining." "Defending the interests of the worker" is a slogan which includes many concerns. The picture may be clarified by the ranks assigned to the four stated union functions. As hypothesized, in OLDS, the factory with the most complex technology, three-quarters thought the most important union function was to fight for better working conditions. In all other factories, working conditions was ranked second: FIAT and IKA, 40 percent; PAL, 33 percent. FIAT and IKA employees thought the most important union goal was to increase wages, but surprisingly, where technology was least complex (IKA and PAL), workers ranked the achievement of social solidarity as second in importance (see Table 8.3). Perhaps, where unions are

TABLE 8.3 MOST AND LEAST IMPORTANT FUNCTIONS OF THE UNION (PERCENTS)

	OLDS		*FIAT*		*IKA*		*PAL*	
	Importance		*Importance*		*Importance*		*Importance*	
Functions	*Most*	*Least*	*Most*	*Least*	*Most*	*Least*	*Most*	*Least*
Secure higher wages	7	12	46	11	37	11	25	15
Obtain better working conditions	75	0	31	5	25	8	23	14
Promote social unity of the workers	15	5	17	17	32	10	36	11
Change political and social system[a]	2	83	6	68	6	72	16	60
Totals	99	100	100	101	100	101	100	100

[a] If first and second in importance are combined, the results were: OLDS, 4 percent; FIAT, 14 percent; IKA, 15 percent; and PAL, 29 percent.

new and inexperienced, workers recognize the need for social solidarity. The vast majority of employees in all factories ranked political functions of unions as least important. Only three of the 1,189 indicated in the two open questions that political functions were important. In the forced-choice question, less than a tenth in three of the plants ranked politics as most important while six-tenths or more considered it least important. Finally, as hypothesized, the less technologically developed the society, the more workers considered the political function of the union to be important.

To obtain concrete information on union tasks, workers were asked to name the most important issues their local unions should resolve. The data support the hypothesis that the lower the plant's technology, the more workers wanted their unions to pay attention to economic issues, and conversely, the more complex the technology, the more they stressed working conditions (see Table 8.4). Thus, nine-tenths of PAL employees, in contrast to three-tenths at OLDS, named economic functions while 6 percent of PAL and 33 percent of OLDS named working conditions. On the economic front, OLDS and IKA employees were concerned with security because of threatening unemployment, but FIAT and PAL workers, who had experienced little unemployment, wanted the unions to fight for higher wages and fringe benfits. The most important economic issues named were wages and fringe benefits, job security and unemployment, and job classification. Job classification is considered an economic category be-

TABLE 8.4 MOST IMPORTANT PROBLEMS LOCAL UNIONS MUST FACE (PERCENTS)

	OLDS	*FIAT*	*IKA*	*PAL*
Free responses				
Wages and fringe benefits	7	42[a]	30	50[b]
Job classification system	2	2	—	37
Security and unemployment	20	—	29	1
Working conditions	33	42[c]	25	6
Internal union problems	27[d]	6	—	—
Other problems	11	8	15	6
Totals	100	100	99	100
Responses to officers' list				
Wages and payments	7	—	41	—
Job classification system	—	15	—	73
Fringe benefits	36	28[e]	22	11
Pace of work	23	33	17	—
Shorter work day and work week	15	14	14	—
Seniority and transfer	19	—	5	16
Reformation of industrial relations system	—	10	—	—
Totals	100	100	99	100

[a] Includes payments for vacations and holidays (32 percent).
[b] Includes 34 percent for a bonus.
[c] Includes working hours and lunch period, 25 percent.
[d] Includes anti-union responses (9 percent), and poor administration (9 percent).
[e] Includes bonus for collaboration (25 percent).

cause most workers who thought they were misclassified felt that they should be receiving higher wages.

Earlier we had asked union officers to name the most important local issues they were facing. We presented their list to the members and asked them to select the most important issue. In all four factories, they named economic issues. The less industrial the society, the more they did so; e.g., almost twice the percentage of PAL than of OLDS workers selected economic issues. As predicted, working conditions, as represented by complaints about the pace of work, was selected more frequently by workers in the factories with the most complex technologies (OLDS and FIAT). In brief, the expected relation between technological complexity and workers' priority of union functions was confirmed.

To summarize: whether interview questions were phrased in general or situational terms, whether the questions were of the free-response or the forced-choice type, workers thought of their unions in job-conscious terms. The primary union goals were to improve economic and working conditions. The less complex factory technology was, the more workers thought that their unions should stress economic issues; the more complex, the more they focussed on better working conditions. Relatively little attention was paid to union efforts to promote social solidarity and political action.

Political Ideology, Political Unionism, and Union Militancy

Even though workers may not think about politics when they discuss their work and unions, the ideological stance of their unions may influence their union militancy and political beliefs. I expected those who believed in political unionism to be politically radical because political unionism was defined in radical terms: changing the political and social system of the country. If political unionists are also the most militant and the most radical, PAL's employees should represent the radical-militant extreme; FIAT and IKA, the middle; and OLDS, the conservative-nonmilitant extreme.

In the United States and India, the index of political ideology contained standard items regarding the desirability of decreasing the power of business in government and increasing the power of labor and the working class. Unfortunately, I could not ask workers about their politics in Italy and Argentina and still get management's cooperation in the research. Therefore, in Italy I decided to use the union with which the worker most closely identified as the indicator of political ideology. Anti-unionists and the politically apathetic were classified as conservatives, SIDA as neutral, UIL and CISL as liberal, and CGIL as radical. When union identification was not available, ideology was determined by scrutinizing the entire interview for the national problems workers defined as most urgent and for their attitudes toward business's political influence. In IKA, the political ideology of workers was determined by the extent to which they supported the union in national politics. Since the union's allegiance to Peronism was a threat to the conservative government in power (Pendle, 1963:142-148), and since labor was the major opponent of the government, support for union politics was a measure of radicalism. More strictly, the index measures support for a laboristic political ideology.

The indicators of political ideology in the four societies are not phenomenological or empirical equivalents. Not only is the political spectrum wider in some countries than in others, but opposition to business domination is structured differently in each nation. In the United States, radicals are rare and liberals are opposed not to business as such, but to its disproportionate influence in government. Although some workers in Argentina were socialists, they could best express their politics indirectly by taking a strong anti-business stance within labor circles. Finally, since India has an avowed socialist government, workers were not asked whether they believed in socialism but whether they felt that the government was doing all it could to help workers and lower classes rather than business and higher-income groups.

A direct comparison of the percentage of radicals, liberals, neutrals, and conservatives in each country is fruitless because the categories are not equivalent and the cut-off points are not the same. However, the two extreme categories may be combined to provide some clue. FIAT workers appear to be the most liberal-radical, followed by PAL's; 67 and 45 percent respectively. Almost half of IKA and OLDS employees fell in the neutral category, but OLDS had more conservatives (see Table 8.5). In extent of radicalism, the samples ranked as follows: FIAT, PAL, IKA, and OLDS.

TABLE 8.5 POLITICAL IDEOLOGY (PI) OF AUTOMOBILE WORKERS AND PERCENTAGE WHO BELIEVED IN POLITICAL UNIONISM (PU) ACCORDING TO THEIR POLITICAL IDEOLOGY

Political Ideology[a]	*OLDS*		*FIAT*		*IKA*		*PAL*	
	PI	*PU*[b]	*PI*	*PU*[b]	*PI*	*PU*[b]	*PI*	*PU*[b]
Conservative	25	—	13	6	19	2	29	30
Neutral	48	5	20	9	48	7	26	31
Liberal	27	4	44	11	18	14	23	25
Radical	—	—	23	29	15	57	22	31
Total	100	3	100	14	100	15	100	29
p of χ^2	$<.20$		$<.01$		$<.001$		$<.50$	
Coefficient of contingency (corrected)	—	—	.319		.622		—	—

[a] The conservative category for FIAT refers to those who were apathetic or hostile toward unions and the neutral category, the politically conservative. For PAL, the conservative and neutral categories should read very conservative and conservative, respectively.

[b] The percent who ranked political unionism as the first or second most important union function.

Political unionism and political ideology were not related in the same way in all nations. In OLDS, too few workers believed in political unionism to permit a reliable comparison; in FIAT and IKA, the two indicators were significantly related, and in PAL they were not related. Since Americans do not believe in political unionism, the question about its relationship to ideology is almost meaningless. Both the Italians and the Argentinians understand political unionism and relate it to their political beliefs. The Indians appear to be too distant from their unions to relate political unionism to the political reality.

The political views of the workers did not correspond to their officers'. In OLDS, all UAW officers supported the Democratic party and urged their members to do the same. But only half did so; half thought that the union had too much political influence. In Italy, almost all the officers believed in political unionism, but only 6 percent of the workers went along. Although the Commissione Interna was dominated by the nonradical unions, only a bare majority of conservatives evaluated it positively. In Argentina, national union leaders strongly favored political unionism, but almost a quarter of the workers spontaneously revealed in the interviews that they thought the union was too involved in politics. Undoubtedly many others felt the same way. In India, union officials were more involved in politics than trade union affairs, but about one-quarter of the members felt that the socialist ideology of their officers was meaningless because they were uninterested in the workers and colluded with management to retain power (cf. Sheth, 1968:167). In short, not only did many workers in all four countries disagree with union officers on politics, but officers were unsuccessful in making the rank and file politically conscious. Up to 30 percent of the workers in each country could not identify a single national issue and the overwhelming majority who did could not specify what stance the union should take on those issues.

Finally, I examined the relationship between union militancy, political unionism, and political beliefs. To tap militancy, I asked respondents:

> In order to improve the social and economic conditions of the workers, which of the following phrases best describes union behavior in general?
> Fights militantly
> Works with determination
> Bargains freely with management
> Cooperates with management's suggestions

Data in Table 8.6 show that the most militant unionists were found in PAL, followed by IKA, OLDS, and FIAT. While half of PAL's employees preferred militant unions, half of FIAT's wanted unions to cooperate with management. FIAT's workers had segmented their views on political ideology from notions of

TABLE 8.6 UNION BEHAVIOR TO IMPROVE SOCIAL AND ECONOMIC CONDITIONS OF THE WORKINGMAN (PERCENTS)

Union Behavior	*OLDS*	*FIAT*	*IKA*	*PAL*
Fights militantly	29	18	39	53
Works with determination	26	13	30	13
Bargains freely with management	33	19	15	27
Cooperates with management's suggestions	12	50	16	7
Total	100	100	100	100

how unions should behave in industrial relations. Evidently radical politics were much more tolerable than militant collective bargaining behavior.

Although political ideology and belief in union militancy were statistically related only in IKA, certain trends appeared in the other societies. Conservatives were most militant in both OLDS and IKA, radicals in FIAT and PAL. A positive association between political unionism and union militancy appeared only in OLDS, but in the other countries radicals did not differ from their work-mates. In short, workers were not consistent in their beliefs on political unionism, union militancy, and political ideology. Inconsistencies were most extreme in FIAT and least extreme in PAL. Edelman (1969) has observed that when radical unions become involved in national politics, they often become conservative, and less militant, in union affairs. This observation seemed to hold for FIAT but not for PAL.

Skill Level and Union Involvement

The internal stratification of the working class according to skill level may mean that workers in the different levels hold different beliefs about union goals and politics. Since skill levels are most highly differentiated where technology is most complex (Udy, 1970:101-102), stratification may be greatest there. But everywhere skilled workers are recruited from a background of privilege. They constitute the top stratum in the factory and strive to maintain their advantages by dominating the local union. Since skilled workers can usually achieve their goals through negotiations, they tend to cushion union militancy.

Factors other than skill may affect union involvement, the most important being education and political ideology. Highly educated young workers whose mobility is blocked may become involved in the union to make it more aggressive. And those socialized in radical politics may see the union as an instrument for class action (Portes, 1971). Therefore union involvement in the four plants was analyzed by education and ideology as well as skill. For economy of reporting, the associations of education and ideology to union involvement will be noted only when the associations differ from those of skill level.

In OLDS, skilled workers tended to be older and more conservative than others, but they were not more highly educated.

In FIAT, the skilled were more highly educated, but they were polarized into radical and conservative camps. The skilled in IKA were both the most highly educated and the most radical, while in PAL they tended to be highly educated and moderately conservative, but the pattern differed by age cohorts.

Data for OLDS conformed to theoretical expectations. The skilled were older, had been union members the longest, had the highest interest in union affairs, attended meetings most often, knew the most about union officers and their activities, and evaluated the union highest. They were both the most politically conservative and the most involved in the union. However, in FIAT skill level was not strongly related to union involvement. Though more of the skilled claimed to be interested in unions, fewer were members, attended meetings, voted, knew union officers, or had knowledge about union affairs. Educational achievement was more highly associated with union involvement than was skill level; the highly educated were more interested and knowledgeable about union affairs, but fewer of them voted in elections. Although political ideology was more highly related to union involvement than either skill or education, the relationship was not linear. Thus, more of the radicals were union members—but they were either highly knowledgeable or totally ignorant of union affairs, or they were highly involved or completely uninvolved in union activities. Multi-unionism complicated the task of characterizing union involvement in FIAT because factors associated with high involvement in one union worked differently in another. This problem is examined in detail later.

In IKA, the most highly involved union members were the skilled workers and the radicals. While the highly educated were most knowledgeable about union affairs, they were less active than were the average and more distrustful of union officers. Finally, in PAL, neither skill, education, nor ideology was associated with union involvement. Where tendencies existed, the skilled were more involved than the average member; the educated, less interested; and the radicals, less involved.

With respect to union goals, I expected the skilled to be the most defensive and to press for fringe benefits, the semiskilled to press for better working conditions, and the unskilled to emphasize immediate wage gains (cf. Shostak, 1969:43). No statistically significant and consistent finding appeared in all four samples, but three trends emerged in the free-response question. First, skilled workers, more than others, emphasized "defending the interests of the workers" and monitoring the internal problems of the union; second, unskilled workers stressed economic bargaining or wage increases; and third, except for IKA, the unskilled and the semiskilled were most concerned about working conditions. In all factories, union officers had listed fringe benefits as an important problem and, except for PAL, they also listed work pace. As anticipated, more skilled workers wanted the union to press for fringe benefits, while the less skilled were more concerned about working conditions in general and speed-up in particular. IKA was the exception because skilled workers there were also concerned with work pace. Their concern was a response to management which sometimes employed them as machine tenders to make identical pieces at a fixed schedule.

I hypothesized that the semiskilled would be most sympathetic to political unionism; the skilled, most opposed; and the unskilled, most apathetic. Too few workers spontaneously mentioned political unionism in the free-response questions to permit a statistical analysis. In the forced-choice question, where political unionism had to be ranked against other goals (raising wages, bettering working conditions, and increasing social solidarity), no trends appeared except in IKA, where more of the skilled endorsed political unionism. To increase their numbers, those who selected political unionism as their first or second choice were combined, but again no difference appeared by skill level. Assembly-line workers and operators of automatic machines are said to endorse political unionism because, having no job autonomy, they become alienated from their work (Lipsitz, 1964; Kornhauser, 1965:224-235). A weak confirming trend was found in OLDS and PAL, but not in FIAT and IKA.

Some scholars (Sayles, 1963; Leggett, 1964) speculate that relations to technology determine union militancy more than political ideology does. Thus, the unskilled avoid militancy, the semiskilled are most militant because they are most exposed to technological changes, and the skilled emphasize orderly collective bargaining because this tactic works well for workers who are in short supply. The more complex the plant's technology, the more the data support these ideas. Thus, skill level was not associated with workers' preference for union militancy in PAL, tendencies in support of the theory appeared in FIAT and IKA, and the theory was clearly supported in OLDS. These findings point to increasing differentiation of the working class with industrialization.

Although the association between skill level and union involvement was stronger in societies with the most complex technologies, the evidence was not always consistent. In all four factories the skilled: (1) tended to be the most active in the union, (2) were least anxious for the union to attack problems associated with technology and working conditions, (3) were most anxious for the union to defend their gains and push for fringe benefits, and (4) were inclined to favor orderly bargaining rather than militant tactics.

Skill and Control of the Local Union

Union officers are more preoccupied with political action than the rank and file, both in the United States, where political unionism is not traditional, and in other countries, where it is. Where officers have never been factory workers, as is the case in many undeveloped countries, their unconcern over daily shop problems is understandable. But where officers have been recruited from the ranks, political involvement must be explained. The higher they are in the labor hierarchy, the more officers lose familiarity with shop floor problems, the more they live in an organizational environment, and the more they realize that economic gains (social security, medical care, taxes) are made in the political arena as well as across the bargaining table. Legislative defeats either radicalize them or make them more aware of the political strength of propertied interests.

Obviously, officers would be more politically effective if they could get the rank and file interested in politics. Limited personal contact with workers forces them to resort increasingly to the printed word. But political attitudes and behavior are forged in daily face-to-face interactions. Those who dominate political communication in the factory dominate local union politics. This study hypothesizes that skilled workers dominate political communication in the plant in a way which may oppose the political goals of officers and the interests of less skilled workers. Why should this be the case?

Skilled workers have more education, training, and knowledge than the unskilled and hence tend to be better informed politically. The skilled also have a broader, more impersonal, and more instrumental view of politics. Since they see politics in occupational rather than in personal terms, they discuss politics with their fellow workers more than with family members or friends. The higher status of the skilled encourages the less skilled to solicit or defer to their opinions. Skilled workers are older than the average worker and more experienced in bargaining over shop issues; they have a more flexible view of politics and are more tolerant of divergent opinions. Finally, the skilled have more job autonomy, more freedom to move about at work and more opportunity to discuss factory and union politics. In the community they participate in more organizations and thus have more opportunity to test and spread their views. In short, compared to other workers, the skilled have more contacts in the plant, union, and community, and these contacts give them an advantage in political communication inside and outside the factory.

During the interviews, workers were asked whether they discussed political and economic issues, what topics they discussed, with whom they discussed politics, whose opinions they valued the most, whether others solicited their views, and related questions. Their answers provide a detailed view of their political communication in and out of the plant. These data were then supplemented with data on the extent of worker involvement in union and national politics, ideological beliefs, tenacity of such beliefs, class identification, and participation in community associations. To simplify presentation, the political communication of workers will be described for each factory, and then the common patterns.

The data for OLDS conformed exactly to the theory: 14 of the 15 items dealing with political communication conformed to the expected pattern. Compared to other workers, more of the skilled engaged in discussions of political and economic affairs and more talked to fellow workers rather than family members and friends. More of the skilled valued the opinions of their fellow workers over those of others and more were tolerant of divergent political views. The skilled knew more about union affairs and participated more in union, church, and other community and national organizations. Finally, the skilled were the most active in local politics, but they most opposed the union's political position and activities. More of the skilled identified themselves as politically independent and as members of the middle class. Predictably, they also had a lower sense of powerlessness and anomie.

I then examined whether extent of union involvement or the political ideology of workers affected their patterns of political communication. Almost without exception, the patterns of political communication and social-system participation of the conservatives and those most active in the union were the same as those of the skilled workers. In short, the skilled, conservative, highly educated workers who communicated most with fellow workers in the plant were also the most active in the union, most involved in local politics, and most opposed to the politics of union officers. Put differently, though the politics of union officers were more strongly supported by the less skilled, few of them communicated their ideas and engaged in politics inside or outside the factory.

The pattern in FIAT was not as clear as OLDS', though it bore some similarities. Thus, more of the skilled in FIAT discussed political and economic events and were more involved in national affairs. They discussed politics with their fellow workers more than with family members and neighbors, and more were sympathetic toward unions. The sample contained too few skilled workers to permit a separate analysis of their political communication in the factory according to political ideology, but it is clear that the conservatives and neutrals (backers of the unions in power), rather than the radicals, were the strongest union supporters, the most militant unionists, the most concerned about internal union problems, and the most insistent that their unions stress economic rather than political goals. In contrast, the radicals were less certain about voting for their union (CGIL) and what policies it should pursue. In brief, while the proposed theory was not strongly supported in FIAT, it seemed to hold for workers sympathetic to the conservative and neutral unions in power (SIDA and UIL).

Political communication patterns were clearest in IKA primarily because the skilled were the most homogeneous stratum in the plant; they were the best educated, the highest union participators, and the most radical. The skilled engaged in more political and economic discussions than other workers did, and held them with their work-mates rather than family members or friends. They had lower than average ideological-tenacity scores and higher than average solicitation of their opinions. Moreover, the skilled had more contacts with their fellow workers inside and outside the plant and were the most active in union, community, and national organizations. Paradoxically, while the skilled were the most advantaged in income, education, and prestige, they were the most radical, had the highest anomie scores, and identified most with the middle rather than the working-class.[1]

The IKA case fits our theory in the sense that the skilled were the highest communicators in the plant and the most active in the union and other organizations. The fact that they were radical rather than conservative must be understood within the Argentinian context. Peronistas represented the main opposition to the business government in power in 1966. Militant unionism was a "radical" movement of the working class to gain and consolidate power in the polity, and in this movement the skilled were the vanguard. They aimed not to abolish capitalism or even overturn the stratification system (see Germani, 1966:390), but to establish the legitimacy and power of labor within the political system. As aristocrats of labor, the skilled identified with the middle class and wanted the labor movement to improve their economic position. The most militant unionists in IKA were the most critical of the union's internal political problems; they wanted the union to fight for *economic*, not political, issues. Like the American skilled workers who fought for union recognition in the late nineteenth and early twentieth centuries (Ware, 1935:176-193), the Argentinian skilled workers were fighting to realize their middle-class aspirations. Once successful, like the Americans they will probably support the status quo and resist the politics of their officers.[2]

Patterns of political communication and union behavior were least clear in PAL, probably because the cleavage between officers and rank and file was the most severe. Moderately educated skilled workers who had longest tenure conformed most to the expected pattern. Compared to others, they participated more in political and economic discussions, talked to their work-mates

[1] A deviant finding is that the most highly educated talked politics more with their friends than with fellow workers. The highly educated were also most informed about the union but participated in it the least. Because they identified so strongly with the middle class, they felt superior to the work they were doing, and wanted to move into nonmanual occupations (see Chapter 3).

[2] See Zeitlin (1967) for an opposite interpretation of the situation in Cuba before and after Castro.

rather than friends and kin, valued the opinions of their workmates over those of friends and kin, had a higher rate of solicitation of their opinions, held political beliefs least tenaciously, and identified most with the upper middle class. Most importantly, they were conservative in politics, most active in it, and most critical of the politics of union officers.

A number of unskilled, long-tenured, low-caste workers were apathetic or very conservative in their politics, religiously oriented, politically inactive, and identified with the lower class. But a new cohort of young semiskilled workers were highly educated, from higher castes, and politically radical. Their work-group and union participation was lower than average, but they were more highly involved in extra-plant activities. Like the highly educated Argentinians, they felt superior to other manual workers, pessimistic about their chances for mobility, and suspicious of their union leaders. In brief, those who communicated the most in the plant were politically most active, and most opposed to the ideology of their officers. Workers whose political ideology was closest to that of union officers were the least involved in the factory's social system, least involved in the union, most antagonistic to union officials, and least identified with the working class.

The Search for Correlates of Political Ideology

The previous analysis had shown that the political ideology of workers is not related to their normative conceptions of the primary union functions nor to their evaluation of union performance. Since the union did not appear to be a school for political ideology, six other explanations for the radical-conservative ideology of workers were examined.

A. One explanation is that those most rewarded by the system tend to be the conservative, and those least rewarded, the most radical. Thus skilled workers who have the highest pay, security, and control over their work are the most conservative politically (Marx and Engels, 1959; Michels, 1959; Lenin, 1943). Using skill level as the main indicator, I found that in OLDS and PAL skilled workers tended to be conservative, but in FIAT and IKA they tended to be radical. Thus it appears that the stratal explanation of ideology is too simple. Lipset and Bendix (1959) suggest that the social origins of workers may be more important than their present status in determining ideology, and Zeitlin

(1967) suggests that the degree of perceived exploitation is the crucial variable. Both of these views will be examined later.

B. The insecurity explanation of ideology holds that those who suffer most unemployment become radical, while people who have job security become conservative (Leggett, 1963a; Zeitlin, 1967). The hypothesis held for IKA, was inversely but weakly supported in FIAT, was curvilinear for PAL; no trend appeared for OLDS. Seniority, as an indicator of economic security, was weakly associated with political conservatism in OLDS and PAL, but not in IKA and FIAT. Those whose previous work experience was primarily in the industrial sector were slightly more inclined to be radical in IKA and PAL, conservative in FIAT, and undifferentiated in ideology in OLDS.

C. Various ideas relate the occupational mobility patterns of workers to their political ideology. The upwardly mobile are said to become conservative, especially in the United States, because they gratefully identify with those above them (Lopreato, 1967b), or remain radical because they are denied acceptance or because they continued to identify with their class of origin (Goldthorpe *et al.*, 1968). But the downwardly mobile are said to remain conservative because they identify with their class of origin (Wilensky and Edwards, 1959), while the nonmobile sons of industrial workers become radicalized (Lipset and Bendix, 1959). Lopreato (1967b) found that the downwardly mobile in Italy were radical, but Jackman's (1972) re-analysis of his data found that they were not.

None of these ideas which link occupational mobility to political ideology was supported in all four samples. For OLDS and PAL, weak support was found for linking upward mobility with conservative politics, but in FIAT and IKA the upwardly mobile tended to be radical. The results for intergenerational mobility, however measured, were precisely the same in all four countries as those for career mobility.

D. A popular explanation of ideology holds that those reared in rural societies tend to be conservative while the urban-born become liberal or radical (Whyte, 1944). But Kahl (1968:117) suggests that in the city and factory the rural-born resent the changes they have to make, and become radical. In support of this, Leggett (1963a) also reports that uprootedness promotes radicalism. Community of residence is said to relate to ideology: those who live outside the city are more isolated and conservative while urban residents are exposed to politics and become radical (Milbrath, 1965:130).

I used three indicators of socialization: community of socialization, community of residence, and father's occupation (farmer or nonfarmer). Only in OLDS was community of socialization related to political ideology, the rural socialized being more liberal. While weak support for the rural-conservative explanation appeared in FIAT, it did not in IKA and PAL. Community of present residence was not associated with political ideology in any nation, but weak support for the hypothesis that city-dwellers are radical appeared in IKA, an opposite trend appeared in FIAT, and no pattern was noted for OLDS and PAL. Finally, fathers' occupation, as an indicator of socialization, was not related to political ideology anywhere. For OLDS, sons of farmers were slightly overrepresented among the politically neutral, while sons of nonfarmers tended to be liberal. In FIAT, more sons of urban skilled workers appeared among the politically apathetic, while more sons of urban unskilled workers appeared among the radicals. For PAL, more conservatives had fathers with white-collar occupations, while liberals had fewest sons of farmers. No trend was found for IKA. In summary, the findings linking socialization and ideology were weak and inconsistent.

E. The work technology explanation stresses that unskilled, routine, and isolated jobs foster alienation and radicalism (Lipsitz, 1964). Sector satisfaction is sometimes related to political ideology: those who dislike the factory (ex-farmers, ex-white-collar workers) are conservative (Whyte *et al.*, 1955) because they prefer their past work, but those who dislike the job but not the sector (manufacturing) become radical (Blauner, 1964).

Although my findings were not statistically significant, in OLDS and PAL more assembly-line workers were liberal or radical, but in FIAT and IKA they were more conservative. In no factory was a majority of employees found in the liberal or radical categories.[3] But in IKA and FIAT craftsmen were more radical and in PAL they were more conservative.

Two other work-related variables, situs and job satisfaction, are related to ideology. Conservatives with farm backgrounds presumably reject the factory, but radicals accept it because they see their future in it and want to reform it. But with respect to job satisfaction, conservatives are said to be satisfied, radicals

[3] OLDS, 37 percent; PAL, 26 percent; FIAT, 25 percent; IKA, 7 percent.

dissatisfied. The data show no statistically significant relationships or trends between sector or job satisfaction and political ideology.

F. A social-integration explanation of ideology stresses that workers who are integrated into their families, neighborhoods, communities, and nations do not become radicalized (Mayo, 1945). Unattached, rootless workers become militant and radical because the union or party is the only source of political socialization (Portes, 1971). The family-integration index was not associated with ideology in any nation, but in OLDS and FIAT, more of the independents or liberals were highly integrated into kinship networks. Church-attenders are thought to be more conservative than the unchurched; this tended to be the case in all factories except in OLDS, where church-goers were politically liberal. The American social-participation literature stresses that those highly involved in community and national affairs are more conservative and see the society as integrated. Only in IKA were ideology and community participation related, but radicals rather than conservatives were the most involved. In the other factories, the conservatives had slightly higher participation rates.

In OLDS and PAL, the conservatives, as expected, were more highly involved in national issues but the trends were not statistically significant. In IKA and FIAT, the reverse situation obtained; the radicals were much more highly involved. Finally, weak and inconsistent associations were found between perceived societal anomie and ideology in three of the factories. No trend was discernible in OLDS, but in FIAT the politically apathetic had the highest anomie scores; the liberals had the lowest; the radicals were in between. In IKA both radicals and conservatives had higher scores than liberals. And in PAL, the very conservative had the lowest anomie scores, but the conservatives and radicals had the highest.

In summary, the political ideology of autoworkers could not be accounted for by any of the six explanations (cf. Portes and Ross, 1974). Nor was any combination of them useful. None of the eighteen indicators of the six positions was related to ideology in all societies. Only six met the test of statistical significance, but in no case did they hold for two or more cases. Even efforts to locate weak trends were unsuccessful; none of the eighteen indicators exposed the same trend in all societies. However, for eight indicators OLDS and PAL exhibited similar patterns as did FIAT and IKA. Three indicators of occupational mobility showed that conservatives in OLDS and PAL had experienced more upward mobility, but radicals were more upwardly mobile in FIAT and IKA. In the other five cases, radicals and conservatives exhibited inconsistent patterns.

I cannot readily explain the weak and negative findings of this search for correlates of ideology. Portes' early suggestion (1971) that radicalism must be explained in terms of socialization, though patently true, does not constitute an explanation. His later suggestion (1974) to explain radicalism by many factors in a causal-path model is sensible. However, the finding that the strongest path corresponds to a factor called structural blame is a validity check on the meaning of radicalism rather than a causal explanation of it.

We are left with earlier theories which explain radicalism in terms of specific historical and contemporary societal conditions. This literature does not explain why different strata of the working class are more radical in one country than another. Under certain circumstances skilled workers seem to be conservative, and under others, radical. In the United States, irrespective of their background and mobility experience, they seem to be conservative; in some European countries and in some underdeveloped countries, the upwardly mobile appear to be more radical. I am not satisfied with the social-psychological explanation that denial of status to skilled workers by people in higher strata makes the skilled radical (Lipset and Bendix, 1959:67). A sociological explanation should explain how segments of the working-class are articulated into the community and the political economy. The following chapters examine in detail how various strata of the working class are differently integrated into social systems of the broader society and how they link the work situation, the union, and the polity.

Conclusions

Although the job-consciousness of American unions is presumably exceptional, this research supports the view that, irrespective of union ideology or personal politics, workers everywhere emphasize job-conscious unionism. Periodically, some scholars assert that American workers are finally becoming radicalized, but this study confirms traditional findings. To my knowledge, no scholar has doubted that Italian unions are ideological, but this research shows that most FIAT workers, irrespective of their union's ideology, reject political unionism. On the contrary, those most active in the union most firmly supported job-conscious unionism. Both conservatives and radicals were critical of the ideological conflicts which made the Commissione Interna ineffective. Even where "radicals" controlled the union, as in Argentina, they were more concerned with getting political influence than with scuttling the economic system. Identified with the middle class and bent on upward mobility, they wanted the union to pursue economic gains.

Kerr *et al.* (1960:221-223) suggest that perhaps the job-consciousness of unions in the Western capitalist nations reflects the bourgeois origins of their leaders. This explanation should not apply to India, where all the unions and the government are socialistic. While more of PAL's workers were radical and believed in political unionism, they felt that the behavior of union officers did not reflect dedication to socialist principles. Workers who were sheltered by the union favored the union; those who were not were against it. The well educated middle-class workers, though the most radical, were the most disaffected union members because they believed that union officers played favorites in the factory.

Three hypotheses were offered concerning the relevance of technology for union involvement. First, since unions in the same industry confronted similar problems, workers everywhere were expected to respond similarly to some union problems. Second, since the factories differed in technological complexity, workers were expected to respond differently to the way their unions handled technologically related problems. Third, since workers at each skill level confronted different technological problems, workers with the same skills in different factories were expected to respond similarly to some problems.

Although conclusions must be held lightly with only four cases, the hypotheses tended to be supported. Regardless of national industrialization, workers everywhere saw their unions as legitimate. Although all of them were favorably disposed toward unionism, interest and participation in union affairs were relatively low everywhere. Whatever the formal political ideology of their unions, workers everywhere wanted their unions to pursue economic gains and improved working conditions, and to avoid politics. Workers judged unions on these grounds and criticized union officers when they pursued other goals.

To some extent, the goals workers wanted their unions to pursue were influenced by the technology of the factory. The more complex it was, the more workers urged unions to pay special heed to working conditions, such as the pace of work. The less complex, the more they awarded priority to economic objectives. Union militancy was unaffected by the technological complexity of the factory.

Skill level, as an indicator of the worker's relationship to technology, had some bearing on their union goal-preferences. Thus, semiskilled workers, who are most affected by technological change, were most concerned about working conditions and work pace; the unskilled were most concerned about wages; and the skilled most concerned with long-term benefits. Contrary to Marxist theory, neither skill nor job operation was related to beliefs in union militancy or to political unionism. In all four factories, skilled workers tended to be the most active in work-group, union, community, and national affairs. The more complex the plant's technology, the more skilled workers dominated the union and political communication in the factory. Except for IKA, the skilled tended to exercise a conservative political influence on their fellow workers and on the unions. If union officials want to radicalize the rank and file and make them more militant. they must somehow overcome the resistance of self-serving skilled workers (Lenin, 1943).

9

Linking Systems for Working-Class Movements

Both Marx and Durkheim observed over eighty years ago that workers were isolated from the associational life of their societies. Both thought that industry was disrupting traditional social organization and that workers were bearing the costs of social change. Although Marx and Durkheim differed in their Utopian views, both predicted that workers would forge a new society which would reduce their suffering (alienation or anomie). Both predicted that workers would build a new society integrated around new social entities (proletarian states or occupational corporations) which would replace the family as the primary organ of societal integration. Finally, both felt that advanced industrial societies would first achieve this new stage of integration. This chapter examines the extent of worker involvement in social systems beyond the factory and whether such involvement increases with the industrial maturity of nations.

Previous chapters have examined how workers responded to systems directly influenced by technology: viz., the work-group and union. Workers everywhere adapted to their jobs easily, developed strong social ties with their work-mates, and responded favorably towards unions. These uniform responses reflected commonalities in the technologies, work situations, and organizations of the four factories. Differences in responses reflected variations in factory technology and the differences in the technology of work operations. But technology probably does not affect the workers' involvements outside the factory so directly. This chapter examines the workers' involvements in such systems as the family, friendship circles, neighborhood, community, and nation.

Participation in nonlocal social systems is of supreme political importance. If, despite their exposure to the factory and union, workers do not become involved in community and national organizations, an effective working-class movement may not arise. A working-class movement appears when workers are conscious of their identity, when they have created a set of special-interest organizations, and when these organizations influence community and national events. In effect, a working-class movement or party must be formally organized and tied into systems extending beyond the family, neighborhood, and work-place. In Britain, the organizational network of the Labour Party and the working-class movement took nearly one hundred fifty years to emerge (Bauman, 1972). The research question is whether these conditions appear more quickly today in less developed industrial societies.

Scholars disagree on the time it takes factory employees to acquire the mentality and organizational involvements of workers in advanced industrial societies. The industrialism hypothesis holds that workers quickly build and respond to organizations which appear with the factory system. The factory is the training ground for industrial culture: it quickly educates workers to understand the need for linking their jobs and unions to community and national organizations. The result is that an organizational network articulating work and nonwork social systems does in fact quickly appear. If this is correct, workers in the four countries should be similarly involved in nonlocal social systems despite societal differences in social organization and technological complexity.

According to the development hypothesis (Moore, 1965; Bauman, 1972:170; Hawley, 1971:306-315), newly industrializing societies are poorly integrated and organizationally "vacant." Over time, factories and unions link to other organizations, special working-class organizations arise, society becomes more organizationally dense, and workers become better informed and more involved in community and national issues. This process is reflected in workers' becoming more interested in the union, political parties, and community organizations which have national interests. Thus, compared to less industrial societies, workers in more industrial societies should: (1) be less involved in the family and neighborhood, (2) be more involved in formal organizations of the community and nation, (3) engage in more activities in these organizations, (4) participate in more organizations with their fellow workers, and (5) have more knowledge of union, neighborhood, community, and national problems.

The two hypotheses differ in their estimates of the time required for an organizational society to emerge and the time required for workers to become involved in the organizations. The industrialism hypothesis has some apparent validity in that workers are not great joiners even in mature industrial societies.

Even in the dual economies of developing societies one sector may be organizationally barren and the other quite modern (cf. Dore, 1973:415-416). In the modern sector, the organizational contours of an advanced industrial society may quickly appear. Since factory employees live in this sector, they may become as organizationally and issue-involved as do workers in more industrial societies. Therefore I predicted that workers in all four societies would become equally involved in the organizations of their societies, but workers in the more advanced societies might carry over more of their work and union ties into the broader community.

But what difference does it make whether workers become organizationally involved early or late in the industrialization process? The question bears on the appropriate timing for launching a successful working-class movement. Where workers are little disposed to join organizations, political unionism is premature and futile. If union officials try to launch a class party before the workers understand that the resolution of factory and union problems need community and national inputs, the workers will consider the union officials to be self-serving politicians. If, on the other hand, workers become involved in community and national affairs before unions organize them politically, the institutions of pluralistic politics will hamper the formation of a working-class social movement. Political unionism is most successful when workers who are most active in the union are also active in politics and other organizations. If workers who are active in the community are not active in the union, or if they value other organizations more than the union, the opportune period for a class movement may have passed.[1] Although conditions other than work-related community involvements affect working-class movements, organizational readiness is a precondition.

Urban Characteristics and Social Participation

Some attributes of neighborhoods and cities which may affect the organizational participation of workers should be described, as background for interpreting the data to be presented. At the time of the research, the Lansing urban area contained about

[1] Freeman and Showel (1951) found that voluntary organizations have more influence on political behavior than do formal political organizations.

300,000 inhabitants. Although Lansing was the smallest city in the study, its factories had the largest average work-forces. In 1963, they averaged 108 employees (U.S. Bureau of the Census, 1966:23-28) compared to 79 for Turin, and 11 for Córdoba. The government bureaus in Lansing, which employed 23,000 workers, were typically large, and the nearby state university employed 8,500 workers. Large enterprises tend to cluster in communities of high organizational density, a condition which increases the employee's opportunity to participate in organizations and find friends in those organizations.

Although Lansing was not a big city, the area contained many services and special-interest organizations. Clustered in the city center were specialty retail services, business offices, and state government buildings. Most residents did not often visit the city center, because several well-equipped shopping centers were more accessible to residential areas. About half of the labor force of the city and OLDS lived outside the city limits in the open country, villages, and suburbs. Traveling to work and shopping was not costly or time-consuming because four-fifths of the work-force owned automobiles.

Almost all Lansingites lived in single dwellings, and 70 percent owned or were buying their homes. Neighborhoods were relatively small: two-thirds of the families thought them to be under ten city blocks (Smith, Form, and Stone, 1954:277). Most neighborhoods did not have names or distinguishable boundaries. Although a business, grammar school, or church was located within five blocks of most residences, neighborhoods were not self-contained in commercial or institutional services. Neighborhood associations were almost nonexistent. Attending a union, church, or other meeting, or even shopping, usually meant leaving the neighborhood. Yet most residents were satisfied with their neighborhoods, public utilities, and other services. Thus, high neighborhood involvement was an unlikely possibility for most adult males because their friends, services, and organizations were located outside the neighborhood.

In Turin, a major industrial center of 1,000,000 people, FIAT was the city's largest employer but many other large industries were located in the city and the surrounding towns. Since over 1,200 small factories which manufactured a wide range of metal, clothing, and other goods employed fewer than ten workers (Divisione Lavoro e Statistica, 1959:138), typical work organizations were small. Almost 30 percent of FIAT's labor force commuted to work from nearby small towns, mostly by train and bus. Turin's city center contained many cultural facilities, business, and institutional services, but they were not easily accessible except by public transportation.

Most workers, even those residing in small towns, lived in multi-storied buildings with apartments which averaged three rooms. By American standards, neighborhoods (quartieri) were quite large, crowded, and ecologically distinct. Except in the new peripheral sections of the city, neighborhoods had ample shopping facilities and adequate services, such as schools, churches, bars, restaurants, and movies. All neighborhoods had names and official boundaries. Older neighborhoods had strong traditions, giving their residents a strong sense of local identity. Typically, Torinese did not have to leave their neighborhoods to visit friends and relatives, to shop, or to satisfy other needs, but a trip to the factory, union hall, or city center required considerable time and energy. Most Torinese were proud of their city and its traditions.

Córdoba, Argentina's second city with a population of 635,000 in 1967, was the state capital and an educational, food-processing, and service center for a large agricultural region. The city had an attractive central business district with adequate cultural and business services. Although three vehicle-manufacturing factories were situated locally, the average factory in the city employed only eleven workers. Thus most employees were members of small work groups. Almost one-fifth of IKA's employees lived in nearby towns and commuted to work. Compared to most large Latin American cities, Córdoba had few large slum areas, though housing was scarce and seriously overcrowded. About three-quarters of IKA's workers lived in single dwellings; half in scattered areas of the city and the others in the periphery (Dirección General de Estadistica y Censos, 1960:xlv). About half of those who lived in working-class barrios owned their own homes; street paving, water supply, electricity, drainage, markets, schools, and other institutional resources were barely adequate. But in the city's periphery these services were poorer and unevenly developed. About two-thirds of the respondents reported that they did most of their shopping within their neighborhoods. Compared to Turin, Córdoba's neighborhoods were not as self-sufficient in economic and institutional services. Frequently residents had to board dilapidated and crowded buses to shop in the city center or other neighborhoods.

Neighborhood associations were common in Córdoba, but they did not stimulate the exchange of home visits. Workers frequently complained about poor services and utilities, and 40 percent wanted to move to different neighborhoods. Most people left their neighborhoods when they visited bars, attended movies, or witnessed sporting or other events. Importantly, the friends, fellow workers, or relatives with whom they participated in these activities did not reside in their neighborhoods. Thus neighborhood ties were few and fleeting.

Bombay is a megalopolis of 4,000,000 and the center of trade and manufacturing for the western region of India. Although the city is a major manufacturing center, its factories are typically small. In contrast, PAL is a giant enterprise located in Kurla, one of the subcommunities in the urban sprawl of greater Bombay. Only 40 percent of PAL's employees lived in Kurla or adjacent areas; the others commuted to work by bicycle, bus, or train from widely scattered areas. Transportation was so overcrowded and inefficient that workers expected the company to do something about it.

Most residential areas in Bombay were severely overcrowded and lacked public utilities, institutional services, and markets. Although neighborhood associations were common, they proved ineffective in solving local problems. Hence, the majority of workers were dissatisfied with their neighborhoods and wanted to move. Compared to respondents in the other nations, those in PAL had fewest local ties: fewest were born locally, fewest had friends who resided in their neighborhoods, and fewest shopped, visited tea-houses, or attended temples, or organizational meetings in their neighborhoods. Evidence from our interviews did not substantiate the claim (Gist, 1968:27) that residential areas were organized around caste, religion, or village ties. The scarcity of organizations in the neighborhood meant that residents had to leave them for many services.

In sum, Lansing's factories were the largest and most bureaucratic, followed by those of Turin, Córdoba, and Bombay. The larger the factories, the more freedom workers have in choosing friends and joining work-related organizations. The cities followed the same order in their adequacy of public utilities, which are so important for neighborhood satisfaction. Turin's neighborhoods were the most self-sufficient in their institutional and economic services, followed by those of Córdoba, Lansing, and Bombay. The greater availability of services in Turin and Córdoba encouraged neighborhood interaction and involvement, and the absence of services in Lansing and the widespread availability of individual transportation diminished neighborhood participation and involvement. In Bombay, the scarcity of both local services and transportation facilities probably decreased involvement in the neighborhood.

External Influences on Social Participation

Conditions other than the availability of services and transportation can influence participation. Thus, people reared in rural rather than urban communities, especially in new industrial societies, participate less frequently in formal organizations (Freedman and Freedman, 1956); urban and suburban residents participate more than do residents of rural and fringe areas (Wright and Hyman, 1958); parents with school-aged children participate more than childless couples; and people with higher educational achievement have higher participation rates (Hausknecht, 1962).

About one-third of the workers in each sample (except for the more urban IKA sample) was reared on farms, and more of them than the urban-born lived outside the central city. However, community of birth was not associated with social participation in any country. Except for OLDS, present residence (central city or outside) was not related to organizational participation, and in no nation were age of the respondent or the number of children in his family associated with social participation. The percentage of working wives increased with industrialization (from 5 percent for India to 29 percent for the United States), but their employment status had no effect on the husband's participation rate. Finally, OLDS's workers were the most highly educated and FIAT's the lowest, but educational achievement was not related to participation except in PAL. In conclusion, OLDS's employees were the most advantaged in the factors which might increase their social participation (age, number of children, and level of education), FIAT workers were the most disadvantaged (except for age), and IKA and PAL workers most advantaged in terms of urban socialization and residence. Yet these advantages did not influence rates of organizational participation. At another point, I shall examine whether these factors affect involvement in other social systems.

Family Involvement

Sociologists have observed that the social life of manual workers is largely kin-centered (Sussman, 1959), but how much of their free time is spent exclusively with their families is not known. A consistent finding of this study is that workers in all nations did not spend all their free time with their families. About four-fifths reported staying at home or alone during the week in the winter or rainy season, but at least two-fifths reported spending most of their free time in *non-family activities* such as reading, studying, and engaging in hobbies (see Table 9.1). The variety of activities reported (see Table 9.2) is as extensive as that revealed in a San Francisco survey (Bell and Boat, 1957: 392). During weekends and holidays, the percentage of workers who engaged exclusively in home activities dropped precipitous-

Table 9.1 Indicators of Family-centered Activities (percents)

Activities	*OLDS*	*FIAT*	*IKA*	*PAL*
Spends free time alone or with family during week (winter)	73	81[a]	80	81
Engages exclusively in home-based activities[b] during week (winter)	59	47	59	44
Spends free time alone or with family during weekends and holidays	60	58[a]	65	70
Engages exclusively in home-based activities during weekends and holidays	32	37	58	54
Relatives or in-laws live in neighborhood[c]	42	52	49	40
Visited relatives during vacation[c]	8	20	24	N.A.
Would confide in relatives about very personal problems[c]	55	55	53	45
Index of family interaction[c] (0-3), high (2-3)	40	42	45	35
N	(249)	(306)	(275)	(262)

[a] Estimated from the nature of the activities.

[b] Includes watching TV, listening to radio, going to the cinema, doing odd jobs around the house, playing with children, spending time with family, resting, shopping, and taking walks.

[c] Items in the index of family interaction.

ly and non-home activities (participative sports, visiting, attending meetings and planned social events) noticeably increased. This pattern was stronger in the more industrialized societies (see Table 9.1).

Three items in Table 9.1 focus on contacts with relatives outside the nuclear family. About the same proportion of respondents (two-fifths to one-half) in each country had relatives living in their neighborhoods whom they visited about once a week. These figures are similar to those reported by Sussman (1959: 335) for Cleveland, and Rose (1959) for Rome. However, a greater percentage of workers in the less industrial countries reported visiting relatives out of town during their vacations. This finding reflects the fact that more of their relatives still lived in their communities of origin. It may well be that the total extent of kinship visiting actually increases with industrialization simply because a greater percentage of a person's relatives reside in the same city.

TABLE 9.2 SELECTED NON-HOME-BASED FREE-TIME ACTIVITIES (PERCENTS)

Activities	*OLDS*	*FIAT*	*IKA*	*PAL*
Engages in nonfamily activities[a] during week on occasion	37	46	37	41
Engages in nonfamily activities[a] during weekends and holidays	62	63	38	34
Visiting	(18)	(7)	(13)	(16)
Holds a second job	12	11	30	16
Exchanges visits with friends in neighborhood on occasion	88	68	42	79
Occasionally meets work mates outside plant	73	52	72	68
Visits bar, tavern, or tea-house	49	54	46	75
Gets together with friends for TV or radio	36	40	37	35
Attends half or more of the meetings of an organization[c]	22	12	15	24
Attends church or religious meeting once a month or more	46	N.A.	31	75
Attends half or more of the union meetings	8	4	43	42[b]
N	(249)	(306)	(275)	(262)

[a] Includes hobbies, reading, studying, participative sports, attending meetings of organizations, planned social events, and visiting which may be done with the family in or out of the home.

[b] At the plant during lunch period.

[c] Exclusive of union or church.

To gauge the strength of friendship bonds, I asked workers to name people in whom they could confide a secret about a delicate and personal problem. About half named a relative (see Table 9.1), but over 60 percent named fellow workers when asked whether they had such a confidant among their work-mates.[2] When these three questions on kinship relations (have relatives in the neighborhood, visit relatives on vacation, share secret with relative), were combined into an index, about equal proportions (two-fifths) of the workers in the four nations had scores in the upper half of the distribution.

In summary, the limited evidence suggests that the extent of nuclear family involvement in the four nations was similar; about half or more of the workers regularly spent some time outside the home in activities not involving the family. The extent of involvement with kin outside the nuclear family was also similar in all societies, especially when the residence of kin was taken into account. The pattern of family and kin involvement of PAL workers was the most distinctive, probably because more of them were separated from their families, who remained in the villages. Almost two-fifths of PAL's respondents (twice that of other countries) reported spending weekends and holidays in isolation, a figure corroborated by Gist (1968:26), who also reported that two-thirds of the population of Bombay is male.

Workers probably spend more time in the company of their work-mates than with their families, but sociologists disagree on whether worker interaction is intimate or important. The human-relations school (cf. Mayo, 1945), once so popular, stressed the importance of social intimacy in the work group, but recent research (Dubin, 1956; Goldthorpe *et al.*, 1968) has suggested that workers have little emotional investment in the work group. The extensive analysis of interaction in Chapter 5 demonstrated that job contacts were highly conditioned by technological and ecological factors. Where technology was simple and machines occupied little space, workers were crowded close together. The higher the density, the more employees took advantage of interaction opportunities. Operators of complex machines had more autonomy and they initiated more interaction than employees who operated semi-automatic machines. But whether work-group density was high or low and whether worker control over ma-

[2] This figure is strikingly higher than Dotson's (1951); he found that 40 percent of New Haven's couples had no "best friend" outside the family.

chines was high or low, the alleged isolation of autoworkers appeared to be mythical. In all factories, more than half the jobs required some communication, seven-tenths or more of the workers talked to at least three others while on the job, and at least a third talked to eleven or more. Almost six-tenths had two or more good friends in their work group, and an even higher percentage had other good friends in the factory. An equal proportion had close friends at work in whom they could confide intimate secrets. In brief, even under ecological and technological work conditions which minimize interaction, employees everywhere had sufficient opportunity to form strong and lasting friendships.

Workers need not see one another outside the factory to be close friends. On the contrary, a socially supportive work-group reduces the necessity for outside contacts. Although Durkheim (1964:14-18) and others predicted that workers would form a community outside the factory, supportive evidence is not strong (cf. Meissner, 1971). High union involvement should promote the formation of a solidary community, but in all four cases, though workers evaluated their unions highly, their interest and participation in union affairs was low. For the 10 to 20 percent who had second jobs (more in the less industrialized countries), contacts with work mates off the job were minimal (see Table 9.2). For the others, such contacts were sporadic but important. Over two-thirds (half for FIAT) occasionally met their work mates socially, one-tenth reported seeing them socially at least once a week, and almost one-fifth spent some of their vacation together. These contacts probably were intimate. But in addition, approximately one-quarter attended organizational meetings (not the union) with their work mates, an equal percentage discussed political and economic events more frequently with work mates than with others, half or more (U.S. was lower) traveled to work with a work mate, and an equal percentage exchanged visits if they were neighbors (cf. Bell and Boat, 1957:395) (see Table 9.3).

In summary, these data point to a considerable amount of social contact among most workers both inside and outside the factory. Multiple social bonds were the rule rather than the exception, and perhaps a quarter or more of the respondents participated in tightly knit, multiple-bonded primary groups. Nowhere did the nuclear family so completely absorb the worker as to make close friendships with nonkinsmen impossible. Somewhat fewer of the Italians and Indians formed such friendships, but perhaps for different reasons. Distrust of outsiders may be culturally normative for the Italians (see Almond and Verba, 1965:308), and extensive family and kin visiting in the villages by the Indians may have reduced contacts with work mates. Since comparable data on nonwork social bonds are not available for

TABLE 9.3 ACTIVITIES WITH FELLOW WORKERS OUTSIDE THE FACTORY (PERCENTS)

Activities	*OLDS*	*FIAT*	*IKA*	*PAL*
Occasionally meets with work mate outside of plant	73	52	72	68
Exchanged visit with neighbor who is fellow worker[c]	44	58	52	68
Travels or walks to work with fellow worker[c]	36	46	70[b]	50
Fellow worker included in free time activity during week	6	N.A.	6	1
Fellow workers included in week and weekend activities[a]	8	N.A.	30	2
Vacationed with fellow worker[c]	16	7	33	N.A.
Fellow worker is member of common organization, exclusive of union[c]	26	23	20	13
Discusses political and economic issues primarily with fellow worker[c]	33	19	19	24
Index of out-plant interaction[c] (0-5), high (3-5)	21	12	18	11
N	(249)	(306)	(275)	(262)

[a] For first and second activities mentioned occurring during week and weekend holidays.

[b] Fifty-seven percent considered their traveling companions as good friends.

[c] Items included in index of out-plant interaction.

other occupations, it is difficult to know whether autoworkers have more or fewer bonds. I am inclined to the view that sociologists have overestimated the nonwork social ties among white-collar workers (cf. Gerstl, 1961).

INVOLVEMENT IN THE NEIGHBORHOOD

In a highly urban and mobile society like the United States, neighborhood organization may be fragile and unstable, but in newly industrializing societies urban neighborhoods may have the solidarity of transplanted rural villages (Breeze, 1969:Part 3). I tried to gather systematic data on the amount of contacts workers had with friends, relatives, and shopkeepers in the neighborhood. Earlier I pointed out that neighborhood structure may affect interaction. Lansing neighborhoods were small and contained few shopping and institutional facilities; Turin and Córdoba quartieri or barrios were distinct sub-units of the city which contained more facilities; and Bombay's neighborhood contained fewest facilities.

Despite differences in neighborhood facilities, the amount of interaction reported was similar for all the samples (see Table 9.4). About two-thirds reported that they had lived in their

TABLE 9.4 ATTRIBUTES OF NEIGHBORHOOD INVOLVEMENT (PERCENTS)

Attributes	*OLDS*	*FIAT*	*IKA*	*PAL*
Has lived in neighborhood five or more years	66	64	62	65
Prefers remaining in neighborhood	66	71	62	47
Relatives or in-laws live in neighborhood	42	52	49	40
Families of four or more friends and acquaintances live in neighborhood	44	65	70	78
Exchanges visits with friends in neighborhood	91	68	41	79
Most friends visited live in neighborhood	9	32	35	21
Shops more frequently inside than outside the neighborhood	35	77	64	58
Patronizes bar or tea-house more inside than outside neighborhood[a]	31	45	15	34
Identifies neighborhood problems	64	72	89	73
Disposed to attend meetings to discuss neighborhood problems	85	39[b]	77	87
N	(249)	(306)	(275)	(262)

[a] Percent who frequent bar or tea-house are: 47, 53, 46, 74.

[b] This percentage is lower than the others because respondents were asked whether they were disposed to "participate" in rather than "attend" meetings. The results are similar to Rose's (1959:42) finding for two Roman lower class neighborhoods.

neighborhoods for five or more years and wanted to continue living there. The exception was India, where half wanted to move. An equal proportion in each city reported that relatives resided in their neighborhoods and that they visited them about once a week. The less developed the society, the higher the proportion (44 to 78 percent) reporting that four or more of their friends resided close by and that they exchanged visits. However, the great majority in all cities conformed to the urban model in reporting that most of their friends resided in other neighborhoods (see Table 9.4). Unlike the folk-urban pattern where most contacts are local in origin, most workers met their friends in a variety of places—home town, the factory, previous neighborhoods, and in religious, educational, and recreational organizations.

Neighborhood stores, gasoline stations, barbershops, and taverns may serve as focal points of neighborhood interaction (Stone, 1954). Except for Lansing, the great majority of respondents reported that their families shopped more often inside than outside the neighborhood.[3] Tearooms or bars typically were not local social centers, since only a minority reported patronizing them. As an indicator of neighborhood involvement, respondents were asked to identify neighborhood problems and whether they wanted to do anything about them. Two-thirds or more identified problems and even more were disposed to attend meetings to discuss their resolution.

I predicted that neighborhood involvement would be highest in Turin, followed by Córdoba, Lansing, and Bombay, but it was remarkably similar in all four cities. Workers had lived in their neighborhoods long enough to develop close friendships; they regularly visited relatives who lived nearby; they shopped inside the neighborhood, knew its problems, and were disposed to confront them. But unlike the folk-urban pattern, most friendly visiting took place outside the neighborhood. Local involvement was somewhat higher in Turin because its neighborhoods were more self-sufficient in shopping and institutional services, but more of Lansing's residents exchanged visits with their neighbors. In brief, neighborhood involvement was relatively high in all cities and conformed to the expected pattern for urban dwellers (cf. Thomlinson, 1969:181-198).

Community Involvement

Community involvement should be higher for people who were born locally and who reside in the central city. Differences were relatively small: about half of the workers in each

[3] The proportion who shopped in the neighborhood approximates the American pattern except for Lansing. Foley (1950:244) reported that 69 percent of the residents of St. Louis shopped within their local districts.

sample were born in or near the central city, and about three-quarters were born in the state or province. Relatively more of the Indians were born outside the state, and slightly more of the Italians and Argentinians were second-generation urban-dwellers. The proportions living outside the central city increased with the extent of national industrialization. Yet wherever their origin or present residence, the great majority of respondents reported that they liked their city (see Table 9.5). Compared to workers in other industrial countries (cf. Curtis, 1971; Hyman and Wright, 1971), employees in all four factories were highly involved in their communities. Almost all of them read local newspapers regularly and could identify important community problems. Although voting and religious participation are not sensitive crossnational indicators of community involvement, rates of participation were relatively high in all nations. Finally, interest in *both* local and national news was relatively high everywhere except in India, where it was conspicuously lower (see Table 9.5).

TABLE 9.5 INDICATORS OF COMMUNITY INVOLVEMENT (PERCENTS)

Indicators	*OLDS*	*FIAT*	*IKA*	*PAL*
Central city is community of residence	43	72	85	90
Born in city or immediate environs	41	48	49	11
Born in state or province	76	70	74	59
Community of birth and residence the same	23	38[a]	50	16
Likes living in community of residence	91	92	94	83
Voted in last community election	55	98	N.A.	68
Identified one or more community problems[c]	74	69	91	74
More interested in local than national news	40	41	21	22
Interest in both local and national news[c]	38	30	52	8
Belongs to one or more organizations[b, c]	43	33	36	34
Belongs to two or more organizations	15	4	5	8
Active participator in one or more organizations	24	17	15	24
Attends church or religious meeting regularly	46	N.A.	31	75
Index of community involvement (0-3), high (3)	30	21	27	4
N	(249)	(306)	(275)	(262)

[a] Estimated.

[b] Excludes union and church.

[c] Items form the index of community involvement.

Organizational participation is the most general indicator of community involvement. In all nations, one-third or more reported belonging to one or more organizations exclusive of the union and church, but fewer than one-quarter actively participated in them.[4] IKA's employees were the most aware of community problems; the Americans were most concerned with recreational and educational facilities; the Italians and Argentinians with transportation, housing, and unemployment; and the Indians with public utilities. An index of community involvement was constructed of items dealing with the identification of community problems, interest in local and national news, and organizational participation. In accord with the development hypothesis, OLDS's employees were most involved in the city, PAL's the least, and IKA's and FIAT's were in between. However, the most important conclusion to be drawn from the data is that community participation was relatively high for urban manual workers, whatever the economic development of their society.

National Involvement

Voting, acquiring knowledge about national problems, discussing politics and economic issues, and belonging to organizations with national concerns are typical ways in which people get involved with the nation. Voting was not used to measure national involvement because it is quasi-compulsory in Italy and elections had been cancelled in Argentina. The reading of newspapers and periodicals is a better indicator of national involvement. Newspaper-reading was widespread in all four samples, but somewhat lower in India. About two-thirds reported reading periodicals which carried some national news, but again the proportion was smaller for the Indians. Despite the wide exposure to national news, only a minority was more interested in it than in local news —except for India, where the great majority was more interested in national news. Contrary to expectations, more of the workers

[4] Three-quarters of FIAT and IKA workers and one-quarter of OLDS and PAL employees who belonged to organizations belonged to sports or recreational associations. For religious organizations, the figures were: 41 percent for OLDS, 17 percent for FIAT, 1 percent for IKA, and 14 percent for PAL. Twenty-eight percent of PAL's organizational members and 14 percent of IKA's belonged to village or neighborhood associations.

in the less developed societies (Argentina and India) were able to identify and discuss national problems. This probably reflects the massiveness and gravity of problems of those societies as much as the personal involvement of workers in national affairs. Thus, almost every Indian named the war with China and threatening famine as important national problems, but almost one-quarter of the Americans were unaware of any national problem or thought that none existed.

Over half of the Americans were worried about international relations and about a fifth of the Italians and Argentinians worried about government corruption and political factionalism. Cost of living, wage levels, inflation, and unemployment are the traditional concerns of industrial workers. As predicted, the lower the economic development of the society, the greater was the awareness of these economic problems. Only one-tenth of OLDS's workers named these problems, in contrast to over half of IKA's and PAL's. Despite the greater awareness of national economic and political problems in the less developed countries, fewer workers discussed them with other people. In short, although workers everywhere were moderately involved in national affairs, those in the poorer countries were more involved, probably because the problems were more severe. However, the fact that fewer workers discussed problems and fewer belonged to organizations which might resolve them suggest that they appreciated how national events affected their lives, but saw these events as beyond their control. This pattern is typical of factory workers, especially the unskilled, in all industrial countries (Milbrath, 1965:110-141).

An overview of the findings suggests that, with some important exceptions, workers were involved in various social systems from the family to the nation at remarkably similar rates. On this basis, we should expect workers in the four samples to perceive roughly the same amount of societal anomie. On the basis of developmental theory, we should expect workers in the older industrial societies to perceive the least anomie since these societies have had the most time to integrate their institutions. Workers in OLDS and IKA perceived the least amount of societal anomie, and those in FIAT and PAL, the most. The low OLDS scores and the high PAL scores conform to the expectations of development theory, but the lowest scores for IKA are unexpected. Although the explanations for these findings are amplified in Chapter 11,

TABLE 9.6 INDICATORS OF ULTRA-COMMUNITY INVOLVEMENTS

Indicators	*OLDS*	*FIAT*	*IKA*	*PAL*
Reads one or more magazines	69	60	81	48
More interested in national than local news[b]	16	23	18	68
Identified one or more national problems[b]	75	68	88	95
Discusses economic and political issues[b]	57	40[a]	55	30
Voted in the last national election	78	98	N.A.	67
Index of national involvement (0-3), high (3)	32	20	36	51
Median anomie scale score	2.9	3.7	2.0	4.7
N	(249)	(306)	(275)	(262)

[a] Probably should be higher because the question was approached indirectly for Italy and directly for the other nations.

[b] Items included in the Index of national involvement.

it may suffice here to suggest that IKA scores perhaps reflect the expanding industrial opportunities in Argentina; the high FIAT scores, the traditional Italian distrust of organizations (Bonazzi, 1964:60; Almond and Verba, 1965:308); and the high PAL scores, the pessimism of living in a polyglot society struggling to resolve seemingly insoluble problems.

DISCUSSION

These data suggest that industrialization quickly affects the social system involvements of workers. In no society did the family completely absorb their free time as it allegedly does in traditional societies. In fact, workers spent considerable time with their work mates, fellow unionists, and friends from other neighborhoods and organizations. They seemed reasonably well acquainted with the problems of the neighborhood, community, and nation. The evidence did not suggest that increasing industrialization and increasing societal involvement go together (cf. Lerner, 1968:392; Coleman, 1968:400); on the contrary, in all societies only a minority of workers was organizationally active.

Both Marx (1956:185,231-240) and Durkheim (1947:24-31) thought that in mature industrial societies workers would become heavily involved in politics and organizational life. While this has not happened, it may be that the minority who are increasingly link more social systems; that is, those who are most active in work groups become most active in the union (Dean, 1954), the neighborhood, community (Form and Dansereau, 1957), and the nation (Lipset, 1960:373-377). Thus, it may be that although with increasing industrialization workers as a whole do not become more societally involved, a leadership corps does, and this corps becomes the vanguard of a working-class movement.

Data in all four nations were examined for extent to which activists in one social system were active in the others. Two patterns emerged everywhere: those most involved in the family were also most involved in neighborhood affairs, and those most involved in the union were most involved in national affairs (see Table 9.7). It is not surprising that "localites" exist everywhere who are highly involved solely in the family and neighborhood. But proletarian activists must do more; they should be active not only in the work group and union, they should also be organizationally active in pursuing working-class interests in the neighborhood, community, and nation. That is, they should serve as social-system linking-pins.

This linking pattern was most visible in the United States, the most industrial society. Only in the United States were work-group activists also active in the union, and only in the United States were union activists active in the neighborhood, community, and nation (see Table 9.7). In FIAT and IKA those active in national affairs were also active in the work group and union. Thus the linking function in the United States was played by workers most active in the work group and union, while in Italy and Argentina the linking function was played by those most involved in national affairs. Importantly, the American activists were indistinguishable from their coworkers in political ideology, but in FIAT and IKA, they were more radical than their coworkers. PAL activists linked the fewest social systems: only activists in the union were heavily involved in national affairs. But the radicals in PAL were not involved in the union, though they were relatively more involved in the community and national affairs. In short, these findings support the developmental hypothesis: in the most industrial society, system linkages were most organized around the work-plant and union, and in the least industrial society they were virtually nonexistent.

The ideal working-class activist ranks high on sociability, high on working-class identification, high on liberal or radical politics,

Table 9.7 Intersystem Involvements and Related Variables

	SOCIAL SYSTEM INDICATORS[a]								
Social System Indicators	*Work Group*	*Union*	*Neighborhood*	*Community*	*Nation*	*Sociability*	*Anomie*	*Political Ideology*	*Class Identification*
Family	– F – –	– – –	OFIP	– – – –	– – – –	– – – –	– – – –	– – – –	– – – –
Work Group		O[b]F – –	– – – –	O – – –	– FI –	– – – P	OF – –	– – I	O – I[b]P
Union			O – – –	O[b] – – –	OFIP	O – – P	O – – –	– FI –	– – – –
Neighborhood				O – – P	– F – –	– F – P	– – – –	– – – –	– – – –
Community					– FI[b] –	OFIP	O – – –	– FIP	– – IP
Nation						– FIP	O – – –	– FIP	– – – P
Sociability							– – – –	– FIP	– – – P
Anomie								– – I –	– – – –

[a] O for OLDS, F for FIAT, I for IKA and P for PAL. Letters represent statistically significant associations ($\chi^2 = < .05$).
[b] Associations significant between .05 and .10.

and low on anomie. A nine-item index of sociability was constructed with such items as: having good friends in the work group, neighborhood, and voluntary organizations; being socially active (talking to friends at work, in neighborhood, in a bar), watching TV or listening to radio with friends, and exchanging visits with neighbors. In OLDS, those ranking high on sociability also ranked high on union and community participation, and they were undifferentiated from their work mates in politics, class identification, and anomie. In India, workers with high sociability were more active in all systems except the family. Although they were more radical in politics than their fellow workers, they identified more with the middle than the working class. In Italy and Argentina, those high on sociability ranked high on community and national involvement, were more radical than their peers in politics, and were undifferentiated in class identification (see Table 9.7).

Assuming that those most active in the work group, union, community, and nation are also highest in sociability, are firm in their labor politics, and share the class identification of their work mates, then activists in OLDS were the most effective working-class leaders. But they were less firm in their labor politics than were their work mates because they were more conservative. The Italian and Argentinian activists were moderately effective leaders. They were highly involved in the union and national affairs, had some contacts in the work group, were low in sociability, and tended not to share the class identification of their fellow workers. Indian activists were probably the least effective leaders. They had little work-group contact, high national and union concerns, a high degree of sociability, but they were more radical than their fellow workers in politics and did not share their class identification.

Conclusions

The industrialism hypothesis of social-system involvement was supported by evidence that factory workers everywhere became similarly involved in work group, union, community and national affairs. However, in accord with the development hypothesis, those who were most active in the factory and union and also most active in community and national affairs were found in the most industrial society. In all societies the average worker was sufficiently exposed to nonlocal systems to understand the political necessity for linking them into a working-class social movement. In all societies a small group of activists also arose, but it was not an effective avant garde. In the United States, activists represented a grass-roots stratum, but they lacked ideological fervor. In Italy and Argentina, activists had fervor, but they were not deeply involved in the work group. In India, activists were high on sociability, radical in politics, but organizationally inactive and did not identify with the working class.

In the United States, working-class activists appeared to be absorbed in the system of pluralistic politics; in Italy and Argentina, they seemed more concerned with union politics than building social solidarity in the factory and community, and in India, they talked radical politics, but did little else. It appeared that the opportune time to launch a working-class social movement had already passed in the United States, was problematic in Italy and Argentina, and had not yet appeared in India.

10

Internal Stratification of the Working Class

Some Theoretical Speculations

An apparent anomaly was exposed in the previous chapter: that although workers had forged solidary work groups and had become involved in many social systems outside the factory, the most active ones did not link these systems in a way that would promote a successful working-class movement. Undoubtedly different sociohistorical conditions explain this failure in each society, but one structural feature of class may inhibit its solidarity everywhere: the persistent cleavages between skilled and less skilled workers. In the structural differentiation which attends industrialization (Smelser, 1963:105-116), the skilled, compared to the less skilled, may become more involved in nonfactory social systems, more differentiated in their aspirations, and more independent in their politics (Mackenzie, 1973).

Obviously, there are many reasons why the working class may become stratified. I am concerned here primarily with technological factors and technological changes. Scholars have noted that different technological arrangements in the factory influence the amount of discretionary time workers have on the job. The higher the skill requirements of the job, the fewer the technological constraints on interaction. Thus skills are differentiated in terms of the opportunities they provide to form solidary work groups. But we do not know the extent to which work-group organization, including stratification, carries over into the community.

Wilensky (1962:2-4) summarizes three possible positions on the relationship between plant and community behavior. First, the spillover hypothesis states that those who interact most in the work group become most socially active in the community, while social isolates in the factory remain uninvolved in community life. The compensation hypothesis holds that those most isolated at work become most active on the outside, and vice versa; the no-relation hypothesis holds that plant and outplant activities are independent of each other.

Each hypothesis presupposes a different pattern of institutional linkage and a different view of human nature. The spillover hypothesis assumes that the work-place dominates other institutions and that activities at work carry over into the community. Meissner (1971:243) argues that social skills learned on the job are useful in community organizations. Since isolated workers cannot learn or practice social skills in the factory, they feel clumsy in organizational life, so they withdraw. The compensation hypothesis assumes that people need social interaction; if they fail to get it at work, they seek it in the community. The no-relation hypothesis assumes that institutions as well as role structures are segmented, and what happens at work has no bearing on people's activities in the community.

Sociologists generally support the spillover hypothesis, and I will use it to build an explanation for the increasing stratification of the working class which attends industrialization. The compensation hypothesis is rejected on the grounds that it is psychological and difficult to test. We have no data showing that isolated workers who participate in the community are less frustrated than isolated workers who do not participate. Nor do we know the extent that isolated workers fail to participate in community affairs and remain frustrated. Sociologists also reject the no-relation hypothesis on the grounds that institutions are related and that influence in the economic area spreads to noneconomic realms. The extent of institutional articulation in a society depends on many factors. Articulation is tenuous in newly industrializing societies but it is strengthened with industrial growth. The spillover hypothesis has greatest validity where institutional articulation is the highest and, within those societies, for workers who have the most control over their work (e.g., the skilled). The no-relation hypothesis better describes the situation in less industrial societies, and within those societies, the workers who have least control over their job (the unskilled). According to the spillover theory, the skilled should be most involved in both work and nonwork social systems in all countries, but they should be most involved in the most industrial countries. If influence in the working class accompanies participational advantages, the influence of the skilled over the less skilled should increase with industrialization. Why should this be so?

Industrial societies are consumption societies and industrial workers, like others, seek to maximize their consumption. Social participation and engaging in "creative" leisure are consumption activities which cost money. Among industrial workers, the skilled earn the highest wages, consume the most, participate the most, and influence others the most. The crucial advantage of the skilled is their greater ability to forge cohesive work groups because they have the most job autonomy. The more advanced the technology, the more job autonomy they have, because the amount of discretionary time on the job is greatest where technology is most sophisticated. Thus, in the United States where craftsmen are used as specialists (toolmakers, tool repairers, tool maintainers), they have considerable freedom to move about, but in newly industrial societies, the skilled are often used in production (Davis and Goodman, 1972:27) to make pieces, and they have less job freedom.

The strategic use of time enables the skilled to build solidary work groups which can be easily galvanized in struggles to control the local union (Aronowitz, 1973:46-48). The amorphous work groups of the unskilled limit their political effectiveness. The skilled learn that unions increase their strength by linking to other unions, by working with political parties, and by supporting organizations with similar community and national interests. Their greater involvement in social organizations becomes both a style of consumption and a pattern of exerting influence inside and outside the factory. This pattern solidifies with industrialization; as industrial societies become more organizationally dense and integrated, skilled workers become more involved in both formal and informal systems inside and outside the factory. Since the system involvements of the less skilled change very little, the cleavages within the working class tend to increase.

I offer this theoretical model timidly because it seems to fly in the face of other evidence. Thus, wage differences between the skilled and unskilled decrease with industrialization, a fact the skilled deplore. Perhaps the skilled are fighting a rear-guard action to slow their absorption into the working class, as the Marxists seem to feel. Or perhaps the skilled are advancing an independent political program, as Mackenzie (1973) seems to believe.

From Privileged Origins to Elite Status

Does the social background of the skilled give them an early advantage over other workers? Data in Chapter 3 showed that the skilled were recruited from more privileged backgrounds than other employees. In each nation, more skilled workers had grandfathers in non-farming occupations and fathers who were themselves skilled. They provided their sons with superior educational and vocational training so that they experienced more upward occupational mobility than most industrial workers. Even though the skilled were older than their work mates, they experienced less unemployment and were more satisfied with their jobs. The wages of the skilled averaged 50 percent more than those of the unskilled, in all the companies. Finally, since more of the skilled lived in the central city, they could travel to union and other organizational meetings more readily than the less skilled.

All these advantages were available to the skilled in each factory, except in PAL, where many young, high-caste semiskilled workers had more privileged backgrounds than did older skilled workers. Even some long-tenured unskilled workers earned higher wages than newly hired semiskilled workers (see Chapter 3).

Solidary Work Groups, Union Participation, and Politics

Do the skilled in fact form more solidary work groups than the less skilled, and hence have more opportunity to control the union? Data in Chapter 5 showed that the higher their skill, the more workers were required to communicate with others to do their jobs, the more they moved about freely at work, the more contact they made with others, the more they talked to others during working hours, the more intimate their contacts became, the more confidants they developed in their work group, and the more they liked their work mates (see Table 10.1). In short, the evidence supports the conclusion that skilled workers forge more solidary work groups than the less skilled and that they are most successful in factories with the most complex technologies (OLDS and FIAT).

Blumberg's (1968:217) review of the international literature on union participation concluded that the skilled hold more local offices than the unskilled, including representation on the policy-making workers' councils. The present study also showed that the greater work-group solidarity of the skilled carried over into the union: they had been union members longer, attended meetings more often, were more knowledgeable about the noneconomic

TABLE 10.1 COEFFICIENT OF CONTINGENCY (CORRECTED) BETWEEN SKILL LEVEL AND INDICATORS OF SOCIAL INTERACTION AT WORK

Items	*OLDS*	*FIAT*	*IKA*	*PAL*
Must talk to five or more workers	.344	.306	N.A.	.370
Can move about freely at work	.579	.489	.564	N.S.[c]
Twenty-five or more persons in work space	.408	.546	.209[a]	T[b]
Talks to eleven or more people	.313	.562	.222[a]	.222[a]
Has confidant among co-workers	.270	.163[a]	.228	T[b]
Two or more good friends among co-workers	T[b]	.177[a]	.195[a]	N.S.[c]
Satisfied with contacts with co-workers	.228	N.S.	.077[a]	N.S.[c]
Index of quality of interaction (high)	.441	.292[a]	.374	T[b]

[a] Probability of χ^2 is between .05 and .10.
[b] Trend, but probability of χ^2 above .10.
[c] Not significant by χ^2 test at the .10 level and no trend.

activities of the union, knew more local union officials by name, showed more interest in union affairs, and discussed political and union issues more often with their fellow workers (see Table 10.2). The dominant posture which workers wanted their unions to take toward management varied in each country. In the less industrialized countries of India and Argentina, they wanted their unions to be militant, while in Italy and the United States they preferred orderly bargaining and cooperation with management. Moreover, differences among the skill levels about the extent they preferred unions to be militant increased with industrialization.

Political activity and ideology also varied by country and skill level. In OLDS, the skilled were both more active politically and more conservative than the less skilled. In FIAT, four unions, ranging in ideology from business to communist, competed for the workers' loyalty, but only 15 percent were members and even fewer attended union meetings. The two unions with the most power (SIDA and UIL) were supported by liberals and conservatives. In IKA, the skilled were the most highly involved in union affairs and the strongest supporters of a laboristic political ideology. In India, recently employed, highly educated, semiskilled workers were both the most radical and the least active in union and political affairs. In short, skilled workers were highly involved with unions everywhere, but their involvement was stronger and more conservative in the more industrialized countries.

TABLE 10.2 INDICATORS OF UNION INVOLVEMENT BY SKILL LEVEL (PERCENTS)

Skill Level	*OLDS*	*FIAT*	*IKA*	*PAL*
Attends Meetings Half or More Times				
Unskilled	43	11	37	36
Semiskilled	54	9	31	56
Skilled	73	2	44	43
Total	57	9[a]	36[a]	42[a]
Identifies Names of Union Officers				
Unskilled	41	38	42	65
Semiskilled	48	39	41	57
Skilled	66	28	50	79
Total	51	37[a]	44[a]	66
Opinion of Union is Favorable				
Unskilled	79	74	73	78
Semiskilled	78	84	62	80
Skilled	95	84	73	72
Total	83	81[a]	69[b]	77[b]
Discusses Political & Econ. Issues with Work Mates				
Unskilled	45	17	40	41
Semiskilled	66	27	35	37
Skilled	60	37	67	29
Total	60	26	46	36[a]
Dominant Opinion on Preferred Union Posture				
	Bargain	*Cooperate*	*Militant*	*Militant*
Unskilled	38	41	45	52
Semiskilled	26	52	35	54
Skilled	43	59	34	53
Total	33	50[a]	39[a]	53[a]

[a] Not significant by chi-square test at the .10 level.
[b] χ^2 probability .05 and .10.

WORK-MATE CONTACTS OUTSIDE THE FACTORY

Data on factory and union interaction showed that skilled workers formed a more cohesive occupational community than the less skilled did. Did this cohesion carry over in the community as Durkheim (1902, 1964:64) and Simmel (1922, 1955:185-189) had predicted? Table 10.3 shows that more of the skilled than nonskilled had extra-work contacts with both work mates and friends in other departments. Moreover, more of the skilled saw their work mates in meetings of voluntary organizations (not including the union). In FIAT and IKA, more of the skilled spent some of their vacation with their fellow workers, while in OLDS and PAL they exchanged more visits with neighbors who were fellow workers. Finally, more of the skilled in every nation had more intimate contacts with their fellow workers, and the differences among the skill levels increased with industrialization. (See Table 10.3.) The index of outplant contacts with fellow workers shows higher rates for the skilled in the more industrial societies (cf. Wilensky, 1961b:553).[1]

Free Time and Family Activities

Unskilled workers tend to limit their social life to the family and kin (Berger, 1960:54-79; Hausknecht, 1964:209) and, as a consequence, have a narrow view of the political process and their ability to influence it (Lipset, 1960:115-120; 359-377). The greater political influence of other occupations is traced partly to their broader social networks. The more numerous and varied a stratum's ties to other groups, the better its knowledge of community functions and the greater its chances to exert political influence.

I compared the free-time activities of workers during the week, the weekend, and vacations, and noted the people with whom they interacted and the institutions or organizations in which they participated. During the week the activities of workers with different skills varied little except for the tendency for more of the nonskilled to report that they rested while more of the skilled read books and magazines. During weekends (especially in the United States and Italy) more of the unskilled (especially the assemblers) spent their free time watching sports events and movies and visiting their friends and families, and more of the skilled engaged in hobbies and participative sports. During vaca-

[1] Since single people are freer of family responsibilities than the married and can meet fellow workers outside of work more easily, an analysis of outplant interaction was made, controlling for marital status and skill. No differences were found for OLDS, but in FIAT and IKA, more of the married skilled workers had outplant contacts, although in PAL more of the single skilled workers did. Thus, marital status did not overcome the influence of skill on contacts outside the factory.

TABLE 10.3 FIVE INDICATORS OF INTERACTION WITH WORK MATES OUTSIDE THE FACTORY BY SKILL LEVEL (PERCENTS)[a]

Skill Level	*OLDS*	*FIAT*	*IKA*	*PAL*
Meets Friends in Work Group Outside the Factory				
Unskilled	72	52	77	59
Semiskilled	69	53	62	71
Skilled	84	47	71	75
Total	74	52[c]	71	68[b]
Meets Fellow Workers Not in Work Group Outside the Factory				
Unskilled	62	49	61	75
Semiskilled	58	43	78	76
Skilled	59	30	61	74
Total	59[c]	43[b]	66	75[b]
Meets Fellow Workers in Community Organization				
Unskilled	14	18	15	10
Semiskilled	25	26	19	12
Skilled	57	29	20	17
Total	31	24[b]	17[b]	13
Index of Outplant Interactions (0-5) High (3-6)				
Unskilled	14	29	17	9
Semiskilled	24	37	21	15
Skilled	27	48	14	9
Total	23[b]	36	18[c]	11
Quality of Interaction (0-6) High (3-6)				
Unskilled	33	40	51	65
Semiskilled	63	48	43	65
Skilled	68	56	61	72
Total	59	52[b]	50[b]	68[c]

[a] Unless otherwise noted, the probability of the χ^2s are below the .05 level.
[b] Probability of χ^2 .05-.10.
[c] Probability of the χ^2 above .10.

tions in all countries more of the unskilled stayed in town and visited their relatives. Among those who left town, more of the unskilled visited relatives while more of the skilled traveled to other cities as tourists (see Table 10.4). Undoubtedly the skilled could better afford out-of-town travel, but differences even appeared among workers who stayed in town: more of the unskilled reported doing odd jobs around the house and resting,

TABLE 10.4 PERCENTAGE OF WORKERS WHO SPENT AT LEAST PART OF THEIR VACATION OUT OF TOWN,[a] BY SKILL LEVEL

Skill Level	*OLDS*	*FIAT*	*IKA*[b]
Unskilled	65	24	40
Semiskilled	68	33	43
Skilled	83	52	49
Total	71	34	43

[a] Does not include those workers who went to their home towns to visit relatives.

[b] Probability of χ^2 above the .05 level.

while more of the skilled engaged in hobbies, reading, and participative sports (see Table 10.5). In sum, patterns of free-time activities point to skill differences in life styles; the higher the skill, less passive the activities and the less often they took place in local contexts. Scores on the index of kinship involvement showed no differences by skill for IKA and PAL, but did show differences in OLDS and FIAT, where the skilled were less involved with their kin. Again the behavior of workers with different skills was more unlike in the more industrial countries.

NEIGHBORHOOD INVOLVEMENT

According to theory, compared to the unskilled, the skilled should be less involved in neighborhood activities because they

TABLE 10.5 ACTIVITIES WHICH TOOK PLACE AT HOME AND OUTSIDE THE HOME DURING VACATION[a] BY SKILL LEVEL (PERCENTS)

	OLDS		*FIAT*		*IKA*		*PAL*[b]		
kill Level	*Home*	*Other*	*Home*	*Other*	*Home*	*Other*	*Visit*	*Odd Jobs*	*Hobbies, Reading*
nskilled	44	48	45	35	62	37	43	28	11
emiskilled	30	70	34	45	74	26	26	15	24
killed	14	86	25	54	51	49	30	34	16
otal	29	72	36	43	64	36	33	34	17

[a] Home activities include visiting relatives, doing odd jobs around the house and resting. Other cludes touring and engaging in hobbies.

[b] Since PAL did not have formal vacations, weekend activities are used.

are preoccupied with community and national events. Data in Table 10.6 show that, except for India, despite their longer residence in the neighborhood, the skilled reported that fewer of their friends resided in the neighborhood and that more of the friends with whom they exchanged visits lived in other neighborhoods. Moreover, the longer residence of the skilled did not result in a greater desire to remain in the area. The skilled appeared to be more cosmopolitan, because they had fewer friends who came from their home towns or the same region of the country. Only Italy was an exception, but this is explained by the fact that most skilled workers were born in the Turin area. In Argentina and India, where neighborhood associations were common, fewer of the skilled reported participating in them. Finally, fewer of the skilled shopped and patronized bars or tea-houses in their neighborhoods. Although skill differences in neighborhood involvement were not strong anywhere, when they did appear they

TABLE 10.6 POSITION OF SKILLED WORKERS COMPARED TO LESS SKILLED ON ITEMS RELATING TO NEIGHBORHOOD INVOLVEMENT[a]

Items	*OLDS*	*FIAT*	*IKA*	*PAL*
Years in neighborhood	more*	more	same	more
Prefers to move residence	same	same	more	same
Friends with whom exchange visits scattered in city	more*	more	more*	same
Friends from region of origin	fewer*	fewer*	same	fewer
Friends from home town	same	more	same	fewer*
Percent of friends residing in neighborhood	smaller	smaller*	smaller*	same
Exchanges visits with work mates in neighborhood	same	same	less	more
Participates in neighborhood organization	—	—	less*	less*
Shops outside neighborhood	more	same	more	same
Patronizes bar or tea-house outside neighborhood	more	more*	more	more

[a] Differences between skilled levels below the five percent level by the χ^2 test is indicated by an asterisk. No asterisks means that the χ^2 probability was between .05 and .10. Probabilities above the ten percent level are interpreted as no differences between skill levels.

were in the expected direction. In brief, although the skilled had resided in their neighborhoods longer than the less skilled, the skilled had fewer personal, kinship, and commercial ties; the more industrial the country, the more this was the case.

Community Involvement

The higher their socioeconomic level, the more strata become involved in nonlocal social systems (Milbrath, 1965:101-141; Merton, 1957:402-406).[2] This generalization was expected to hold even though the skilled are only slightly better educated and paid than the unskilled. To be sure, nonstratification factors such as being born locally and being satisfied with the community might affect a worker's community involvement. However, only in Italy were more of the skilled than unskilled born locally and in no country was skill level associated with community satisfaction.

Every indicator of community involvement revealed that the skilled were more involved than the unskilled. In all countries, but especially in Italy and India, more of the skilled subscribed to local newspapers and were interested in nonlocal news and events, and more of the less skilled were interested in purely local news (see Table 10.7). Manual workers participate little in formal organizations (Hyman and Wright, 1971), and this was the case in the four samples. But everywhere, more of the skilled than the less skilled belonged to organizations. Yet the less skilled participated more regularly in the organizations to which they did belong. This unexpected finding primarily reflects different patterns of religious activity; more of the unskilled or semiskilled belonged to religious organizations (excluding the church) and, except for the United States, more of them attended religious services regularly. Meissner (1971:250) regards religious participation as expressive activity. In his study those who had little control over their work (e.g., the unskilled) compensated by participating more heavily in religious organizations. I also found a tendency for the less skilled to participate in organizations with primary-group members; more of them attended religious organizations with relatives and non-religious organizations with friends and relatives, while more of the skilled attended such meetings with fellow workers (see Table 10.7).

[2] Hamilton (1964) feels that despite this, the skilled are more like the unskilled than white-collar workers.

TABLE 10.7 INDICATORS OF LOCAL COMMUNITY INVOLVEMENT ACCORDING TO SKILL LEVEL (PERCENTS)[a]

Item	OLDS				FIAT				IKA				PAL			
	Un	*Ss*	*Sk*	*Tot.*	*Un*	*Ss*	*Sk*	*Tot.*	*Un*	*Ss*	*Sk*	*Tot.*	*Un*	*Ss*	*Sk*	*Tot.*
Interest in local news	50	43	35	42	43	41	37	40	26	19	17	21	20	26	18	22
Reads magazines	67	68	78	70	53	63	61	60	77	84	86	81	22	30	36	29
Organizational member	32	41	67	46	29	35	34	33	34	38	40	37	23	39	45	34
Work mate in organization	14	25	57	31	18	26	29	24	15	19	20	17	10	12	19	13
Religious organization member	26	52	36	41	30	16	6	18	—	—	—	—	69	21	10	14
Attends church weekly[b]	24	28	38	30	N.A.[b]	N.A.	N.A.	N.A.	26	39	28	31	61	60	43	56
Community problems identified																
None and none known	38	21	12	22	39	29	21	31	13	9	4	10	24	26	29	26
Institutional	22	33	31	30	3	10	13	9	3	2	3	3	6	6	14	8
Moral-social	16	13	14	14	32	38	26	34	14	21	13	16	24	23	16	22

[a] In IKA, only 3 persons belonged to religious organizations, but the data for church attendance are for once a month or more frequently. For PAL, data are reported for reading magazines weekly.

[b] Not available.

Finally, respondents were asked to identify the most important problems facing their communities. In all nations except India, the higher their skill, the more workers named specific local problems. More important, they tended to name specific problems which specific community agencies were designed to resolve (e.g., providing health care, educational facilities, welfare and recreational services). The less skilled tended to identify general problems which community agencies could not readily attack, such as the rebelliousness of youth and the breakdown of morality. For most indicators of community involvement, differences among skill levels were larger in the more industrialized countries.

National Involvement

Three indicators of national involvement were used in this analysis: ability to identify national issues which require organizational solutions, participating in discussions of national economic and political problems, and interest in national news. In the previous chapter, I reported that workers in the less industrialized countries identified more national problems probably because their nations were confronting such severe problems that even the least informed were aware of them; war and famine in India and runaway inflation and unemployment in Argentina. However, in conformity with the stratification hypothesis developed here, within each nation the higher the skill, the more workers identified national problems, and this trend was stronger in the more industrialized societies (see Table 10.8).

Table 10.8 Specification of National Problems and Organizational Means to Solve Problems, According to Skill Level (percents)

	OLDS		*FIAT*		*IKA*		*PAL*[a]	
l Level	*Problems Named*	*Means Named*	*Problems Named*	*Means Named*	*Problems Named*	*Means Named*	*Problems Named*	*Means Named*
killed	65	9	61	18	88	7	96	10
iskilled	76	9	67	23	88	6	94	12
led	94	17	85	42	97	15	99	19
al	77	11	68	25	90	8	95	13

The probability of the χ^2 is above .05.

Citizens need little knowledge about national problems to recognize that war and inflation call for government action. They need more knowledge to identify special problems which call for action by specific governmental agencies. Interviews were examined for references to problems requiring action by special governmental agencies such as employment bureaus, medical organizations, educational offices, and economic development agencies. Data in Table 10.8 clearly show that the higher their skill, the more workers referred to specific governmental services, and this trend too was stronger in the more industrialized societies.

On the behavioral level, indicators of national involvement include voting, working for political parties, reading national news, discussing national issues, and attending meetings of organizations which have national concerns. Voting is not a good cross-national indicator of national involvement, but working for political parties directly or through the union is a valid indicator. In no country did more than 3 percent of the respondents engage in such activity, a number too small to analyze by skill level.

Indices of political activity were constructed for the United States and India. The items in the United States index included: interest in politics, voting in local, state, and national elections, contributing to a political campaign, and wearing a campaign button or having a car sticker. As expected, this index showed that skilled workers were more highly involved in politics than the less skilled ($C = .38$). The Indian index of political activity included: interest in political news, voting in community, state, and national elections, membership in political organizations, and discussing politics with a party worker. Although the association between skill level and political activity was weak, more skilled workers read political news and voted in elections. However, the nine most active participants in political organizations were long-tenured semiskilled workers.

Two remaining indicators of national involvement are reading newspapers and discussing political and economic events. In all countries, the higher their skill, the more workers were interested in both community and national news (see Table 10.9). This trend was much stronger in OLDS and FIAT than in IKA and PAL. Finally, more of the skilled, with the possible exception of FIAT, discussed political and economic events. As in the case

TABLE 10.9 INTEREST IN NATIONAL NEWS AND PARTICIPATION IN ECONOMIC AND POLITICAL DISCUSSION BY SKILL LEVEL (PERCENTS)

	OLDS		*FIAT*		*IKA*[a]		*PAL*[a]	
Skill Level	*News*	*Discussion*	*News*	*Discussion*	*News*	*Discussion*	*News*	*Discussion*
Unskilled	50	46	57	36	74	37	80	66
Semiskilled	57	62	59	49	81	47	74	68
Skilled	65	67	63	48	83	62	82	80
Total	58	59	59	45	79	46	78	70

[a] Probability of the χ^2 is above .05.

of the other social systems, differences in national involvement were larger among the skill levels in the more industrialized nations.

ANOMIE

By definition, where social integration is high, normlessness or anomie should be low. People who are highly involved in interrelated systems which extend from the work-place to the nation should see little anomie in their society. Since social-system integration was highest for skilled workers, they should report lowest anomie scores. Also, since system involvement was slightly higher in the more industrial societies, and since these societies are better integrated, anomie scores should decrease with industrialization (cf. Moore, 1965:98-109).

Indian workers had the highest anomie scores, followed by the Italians, Americans, and Argentinians (see Table 10.10). These

TABLE 10.10 HIGH NORMLESS ANOMIE ACCORDING TO SKILL LEVEL (PERCENTS)

Skill Level	*OLDS*	*FIAT*[a]	*IKA*	*PAL*[a]
Unskilled	28	40	13	64
Semiskilled	28	42	9	53
Skilled	8	41	14	60
Total	23	41	14	60
N	(304)	(306)	(315)	(262)

[a] Probability of the χ^2 is above .05.

results generally conform to expectation, except that I expected Argentina to rank second in anomie, instead of lowest. Indian workers were aware of the deep problems of Indian society (see Chapter 9) and the Italians were also disturbed about political conflict, governmental ineffectiveness, and instability. Though the Argentinians were more concerned about national problems than the Americans, they had lower anomie scores. Three local factors may account for this: the majority of the Cordobese were born locally; compared to others in the city, IKA employees received exceptionally good wages; finally, IKA workers were young, well educated, and optimistic about their future in the expanding industrial sector.

Surprisingly, the expected relationship between skill level and anomie was supported only in OLDS, where the skilled had the lowest scores (see Table 10.10). In FIAT and PAL there was no relationship, but in IKA it was reversed. Interpretation is hazardous. Since most of FIAT's employees were newly exposed to industrial and urban life, they may have responded similarly despite differences in skill. In India, problems of the society are so immense that all workers recognize them and believe that they are beyond human control. Although the skilled in IKA had higher anomie scores than the unskilled, their number was small. Only 20 percent of the skilled had high scores compared to twice that for FIAT and three times for PAL. IKA's skilled employees were relatively young, highly educated, upwardly mobile, and highly aspiring. They were dissatisfied with their jobs and planned to quit at the first opportunity. Most did not aspire to better industrial jobs, but wanted to become proprietors, managers, or officials. Why workers who suffer status deprivation see society as anomic will be discussed in the next chapter.

A Partial Reanalysis Using Hierarchial Log–Linear Models[3]

After completing most of this book, Kenneth Spenner (1975) analyzed some of the data with techniques developed by Professor Leo Goodman. Interpreting some of the tables of this chapter with Goodman's techniques enables the researcher to test more precisely the main and interaction effects of the three variables—

[3] With Fred C. Pampel, Jr.

industrialization, skill level, and system involvement. The hypothesis of increasing stratification of system involvements with industrialization specifies a significant three-way interaction effect which may be tested using hierarchial log-linear models developed by Goodman (1970, 1971, 1972). We shall first sketch the logic behind the method and then apply it to a small sample of questions used in this chapter. A definitive test of the hypothesis of the internal stratification of the working class would entail reanalyzing all the data of this book, but we shall select only 25 questions for reanalysis. We are using only one of several ways of ordering the data, but all of them have some limitations (Form, 1975b). Our task is to demonstrate a mode which may be useful to explore the type of hypothesis proposed rather than to replace the previous analysis.

Some of the terms and calculations to be used will be briefly explained before presenting the findings. An effect of some combination of variables accounts for variation in cell frequencies. A model is a set of artificial data generated by an iterative procedure (Goodman, 1972) based on the presence or absence of various effects. Thus, analysis of contingency tables using log-linear models involves creating several models, comparing artificial and actual data, and determining the discrepancy to assess the size of certain effects.

The variation in a three-way table may be decomposed into one-variable effects, two-variable effects and three-variable effects. A single-variable effect—participation (P), industrialization (I), or skill level (S)—is a difference in the cell frequencies which reflects the marginal distribution of one variable. A two-variable effect—(PI), (PS), or (IS)—reflects an association between two variables, or the effect of the joint marginals on the cell frequencies. A three-variable or interaction effect (PIS) exists when a two-variable effect differs depending on the level of the third variable; it is the effect of the three-way marginals on the cell frequencies. Thus, the variation in the cell frequencies is a function of the following effects:

$$(P)\ (I)\ (S)\ (PI)\ (PS)\ (IS)\ (PIS).$$

Calculation of artificial data based on all effects will exactly duplicate the actual cell frequencies. To determine the size of any individual effect, several models which systematically exclude individual effects are specified. The discrepancy between the actual and artificial data indicates the importance of the effect.

The theory suggests which effects need to be tested. In this case, we hypothesize the existence of a significant three-way interaction. Thus, artificial data are calculated for a model excluding (PIS): (P) (I) (S) (PI) (PS) (IS). Note that the method of calculation is based on hierarchial effects. For example, inclusion of the (PI) effect also includes the (P) and (I) effects. The above model can be therefore described in terms of the three pairwise effects (PI) (PS) (IS).

Column 2 shows the chi-square difference between the artificial data and the true data. The difference corresponds to the effect of the interaction since the true data consists of the pairwise effects and the interaction (chi-square) effects. If the chi-square is significant, the model with pairwise effects only is not sufficient to account for the variation—interaction plays an important role.

Columns 4 and 6 present the chi-squares due to skill-participation and industrialization-participation, respectively. A significant chi-square indicates a relationship between participation and skill or industrialization. The chi-squares are obtained by specifying a model which excludes (PIS) and (PS) or (PI). Subtraction of the chi-square due to the interaction alone from the chi-square due to interaction and participation-skill or participation-industrialization gives columns 4 or 6.

To sum up, the chi-squares in Table 10.11 represent the variation explained by interaction, by the relationship between participation and skill, and by the realtionship between participation and industrialization.

A final step is to make the size of the chi-square more meaningful by calculating the percent of association which an effect accounts for (Goodman, 1970). Given the original model,

(P) (I) (S) (PI) (PS) (IS) (PIS),

the theory makes no predictions concerning the relationship between industrialization and skill, nor concerning the single variable effect of participation. Generating artificial data for the model (IS) (P) gives the chi-square, or unexplained variation, relevant to the theory. Thus, we want to find the proportion of relevant association in column 1 that is due to the interaction, skill, and industrialization effects. Dividing the chi-squares in columns 2, 4, and 6 by the chi-square in column 1 gives the percent of relevant association explained by each effect.

Three broad generalizations may be made of the decomposition of the relevant associations in Table 10.11. First, the percentage of the association accounted for by industrialization (column 6) is generally larger than that accounted for by skill alone (column 4) or by the interaction of all three variables (column 2). Moreover, the industrialization effects, which are universal, substantial, and statistically significant, account for almost three-fifths of the total association on the average for each question. Second, skill effects (using .15 level of satistical significance to indicate a trend), were widespread, appearing in almost seven-tenths of the items, and accounting on the average for one-fifth of the total associations. Third, statistically significant three-way interactions between skill, industrialization, and participation appeared for slightly over one-third of the items, accounting for almost one-quarter of the total associations on the average. Three-way interactions tended to be larger than those accounted for by skill alone, but they were smaller than those accounted for by industrialization alone. Briefly, industrialization effects are universal and account for the major proportion of the associations; the skill effects are pervasive and account for about one-fifth of the associations when they appear; and three-way-interaction effects appear in about one-third of the items, and they are moderately strong when they appear.

It is not necessary to comment on the areas of involvement which are affected by extent of national industrialization, because all were affected. Skill effects were also widespread, *but fewer items appeared dealing with union involvement*. We are most interested in three-way-interaction effects because they embody the hypothesis that skill level, extent of industrialization, and participation are associated differently in each country.[4]

The places where the three-way interactions appear may be more important than the number of significant interactions. The questions in Table 10.11 may be classified in three broad areas of worker involvements: with unions and politics, with fellow

[4] The existence of a significant three-way interaction effect does not identify the direction of the interaction. Skill differences may be highest in either the most or least industrialized nation; the size of the interaction effect would be the same in each instance. To interpret the results, the original tables must be consulted.

TABLE 10.11 DECOMPOSITION OF RELEVANT ASSOCIATION FOR TWENTY-FIVE COMPARISONS

Question	(1) Relevant Association Effects χ^{2mL} (P) (IS) (P)	(2) Three-way Interaction Effects χ^{2mL} (b) (PI) (PS) (IS)	(3) Three-way Inter-action: % of Association	(4) Skill Effects χ^2 P	(5) Skill % of Association	(6) Industrialization Effects χ^2 P	(7) Industrialization: % of Association
Unionism: support	26.27 (.006)	12.63 (.049)	48[a]	1.59 (.400)	8	11.80 (.02)	46[a]
Union evaluation	76.81 (.000)	12.63 (.049)	16[a]	1.98 (.350)	3	60.96 (.00)	79[a]
Union posture	40.56 (.000)	13.00 (.043)	32[a]	.79 (.50)	2	26.79 (.00)	66[a]
Attend union meetings	317.61 (.000)	6.00 (.423)	2	1.61 (.4)	1	305.72 (.000)	96[a]
Discuss politics w/ fellow workers	46.71 (.000)	12.64 (.049)	27[a]	13.39 (.001)	28[a]	18.36 (.000)	37[a]
Work mate in same organization	46.88 (.000)	14.08 (.028)	30[a]	4.54 (.120)	10[a]	28.09 (.000)	60[a]
Acts outside home	157.60 (.000)	17.81 (.007)	11[a]	21.21 (.000)	13[a]	114.74 (.000)	73[a]
Vacation w/fw	88.44 (.000)	7.81 (.098)	9[a]	1.36 (.500)	1	76.27 (.000)	86[a]
Know community prob.	589.57 (.000)	80.25 (.000)	14[a]	59.93 (.000)	10[a]	318.29 (.000)	54[a]
Anomie	131.58 (.000)	21.05 (.000)	16[a]	4.32 (.15)	3[a]	106.04 (.000)	80[a]
Org. member	27.70 (.004)	6.57 (.363)	25	14.08 (.000)	50[a]	6.82 (.075)	21[a]
Particip. org.	41.90 (.000)	4.79 (.500)	11	10.54 (.010)	25[a]	31.45 (.000)	75[a]
Union particip.	66.80 (.000)	3.51 (.470)	5	20.10 (.000)	30[a]	46.45 (.000)	67[a]
Index community invol.	50.93 (.000)	2.43 (.500)	5	14.51 (.000)	29[a]	32.70 (.000)	64[a]
Discuss polit. issue	75.43 (.000)	4.65 (.500)	6	14.89 (.000)	20[a]	50.64 (.000)	67[a]
Specifies natl. prob.	204.19 (.000)	4.95 (.500)	2	24.55 (.000)	12[a]	166.95 (.000)	82[a]
Index natl. invol.	95.92 (.000)	7.02 (.318)	7	31.44 (.000)	33[a]	59.50 (.000)	62[a]
Vacation out of town	81.45 (.000)	1.68 (.500)	2	18.65 (.000)	23[a]	60.39 (.000)	74[a]
Quality of interact.	45.53 (.000)	5.57 (.473)	12	17.85 (.000)	37[a]	19.65 (.000)	41[a]
Meets fw outside plant	21.12 (.000)	7.75 (.256)	37	.81 (.500)	4	12.92 (.000)	62[a]
Index soc. inter. fw	75.40 (.000)	2.47 (.500)	3	9.53 (.01)	13[a]	61.60 (.000)	83[a]
Interact. fw outside plant	23.41 (.015)	4.90 (.500)	21	3.59 (.15)	15[a]	14.93 (.005)	64[a]
Neighborhood prob. named	45.23 (.000)	6.76 (.343)	15	1.98 (.4)	4	37.76 (.000)	84[a]
Natl. loc. news inter.	52.43 (.000)	3.98 (.500)	8	4.15 (.15)	8	41.56 (.000)	80[a]
Reads magazines	82.08 (.000)	3.43 (.500)	4	7.21 (.035)	9[a]	75.86 (.000)	91[a]

workers outside the factory, and with community and national affairs. Statistically significant three-way interactions were found primarily in the union-political arena. The first four items in the table deal with union involvement. Support for unionism in general, positive union evaluation, attendance at union meetings, and union posture toward management increased with industrialization. These effects are statistically significant. Ignoring level of industrialization, we find that there is no relationship between the skill composition of the factories and the indicators of union involvement. However, the relationship between skill level and union involvement is different in each of the four countries; that is, there is a significant three-way-interaction effect, as hypothesized. Data in Table 10.2 illustrate this: the more industrial the country, the more skilled workers supported unionism in general, the more positively they evaluated their own union, and the more they wanted the union to bargain cooperatively with management rather than use militant tactics. We should emphasize that the size of the interaction effect is largest for support of unionism in general and for the posture unions should take in collective bargaining.

The fifth and sixth items in Table 10.11 deal with political and organizational involvement. Discussion of political and economic issues with fellow workers, and also fellow workers who are members of the same organization, increases with industrialization. Ignoring the industrial level of the nation, participation in political and economic discussions with fellow workers and participating in common organizations with them increase with skill level. Understandably, the relation between skill and participation for these two items is different in the four countries; there is a significant three-way interaction between skill, participation, and industrialization. In Tables 10.2 and 10.7, we see, especially in IKA and OLDS, that more of the semiskilled and the skilled than the unskilled engage in economic discussions and that in both countries, despite their generally higher levels of participation, the higher the skill, the more workers participated in the same organizations outside the factory.

The different social integration of the workers outside the factory is reflected in their practice of spending some of their vacation time together. The three-way interaction shows that patterns were different in each country. IKA's was most distinctive: more of its employees vacationed together, and the lower their skill, the more they did so.

Next, three-way-interaction effects were apparent in the item dealing with knowledge of community problems. In Table 10.7 we see that in accord with our hypothesis, the skill levels were undifferentiated in India and IKA, while in OLDS and FIAT, the higher the skilled, the more community problems workers named. This greater external orientation of the skilled in the more industrial countries is reflected in the activities outside the home during their vacations. Again, a different pattern of three-way-interaction effects occurred in each country. As the data in Table 10.5 indicate, the higher the nation's industrialization, the greater proportion of workers engaged in non-home activities and the greater the differences among the skilled levels.

Finally, a different pattern of three-way interaction appeared for the item of the anomie workers perceived in their society. As Table 10.10 shows, Italian and Indian workers had higher levels of anomie than the American and especially the Argentinian. The differentiation by skill levels was not great in the less industrial countries, but in the United States the skilled workers had much lower anomie scores than the less skilled.

Since the independent effects of skill were pervasive but not universal, it is important to note where they did not appear. Strangely enough, they were absent in the area of union involvement, where there were strong three-way-interaction effects: attendance at union meetings, approval of unionism, evaluation of union performance, and union posture toward management preferred. Independent skill effects were evident in most of the remaining areas dealing with organizational, community, and national involvements. Since industrialization effects were also operating in these areas, the participational differences were not large enough among the samples to produce statistically significant interaction.

Altogether the data seem to suggest that social or interpersonal relations among workers were less important than union and political solidarity. Thus three-way interactions were conspicuously absent for questions dealing with the amount of social (not organizational) interaction with fellow workers outside the factory: interacting with fellow workers in the neighborhood, meeting fellow workers socially outside the factory, number of friends who are fellow workers, and so on. Three-way interactions were also absent for organizational (church) memberships, reading habits, and related items dealing with community and national involvement. This means, not that the skilled workers in the more industrial countries were not more involved in community and national affairs than the skilled in the less industrial countries, but that national differences were relatively small. From this limited analysis, we may cautiously conclude that the *linking* of union involvement to discussion of political and economic events with fellow workers who are members of the same community organizations underlies the stratification of the skill levels in the more industrial societies.

Conclusions

Factors other than workers' skill may influence their system involvements, the most important being: age, marital status, education, community of socialization or residence, and education. In the previous chapter, I noted that age, marital status, and number of children were not associated with system involvements, but that those socialized in rural communities were more involved in their neighborhoods than the urban-born. In FIAT and OLDS, present residence and system involvement were associated in the expected way: those living in the periphery of the city were more involved with kin and neighborhood, while city dwellers were more involved in community and national systems.[5] In IKA residence had no effect on involvement, but in PAL city residents were more involved in community and national affairs. Finally, in all countries the better educated spent more free time reading and were more interested in national news. When community and national involvement were controlled for education, the association with skill was strongest in OLDS. In short, while variables other than skill were associated with system involvement, none were as strongly and consistently associated as was skill level.

Data from this chapter demonstrate that the labor force of the automobile industry is not a homogeneous mass, but is socially stratified. The system involvements of skilled workers tend to

[5] This pattern was absent in India because more Bombay residents than non-central residents were unskilled and had below average education. In the United States, the superior education of non-city residents did not increase their involvement in city and national affairs.

conform to the spillover hypothesis and those of the less skilled workers to the no-relation hypothesis. Differences among the skill levels in system involvements tend to increase with the level of industrialization of the country. The greater social solidarity of the skilled was shown to be an outgrowth of working in a less restrictive technological environment, and their greater union influence was probably based on their greater solidarity. Moreover, more of the skilled carried their work and union contacts into the community. They were less absorbed in kinship and neighborhood networks and participated more in community affairs. They were more interested in and more knowledgeable about both community and national affairs. In short, compared to the unskilled, the skilled constituted an occupational community; this was most noticeable in the most industrialized countries.

The social significance of the increasing solidarity of the skilled and their differentiation from the less skilled needs thorough exploration. Findings from this study suggest that the probability of launching a successful political movement, insofar as it depends upon worker solidarity, is greater at the early stages of industrialization, when the structural differentiation of the system involvements is not high and when skilled workers are not integrated into systems external to the factory and union. In early industrialization, the strongest bond among workers may be not their occupations but their common identity as factory workers. The subsequent elaboration of the occupational structure seems to affect the system involvements of the skilled workers more than the less skilled.

Most scholars have stressed the homogenizing effects of industrialization (Kerr *et al.*, 1960): wage differentials among the skill levels tend to decrease over time (Reder, 1968:403-414), the working class fails to be absorbed into the middle class (see literature in Goldthorpe *et al.*, 1969:1-29), and industrial workers are part of an undifferentiated middle mass (Westley and Westley, 1971). But this study suggests that the internal stratification of the working class, as revealed by their system involvements, may be increasing. Considerable data already shows that skilled workers feel that their problems differ from those of other workers (Chamberlain and Cullen, 1971:424-426; Mackenzie, 1973). Thus they may not want to join either a working-class movement or middle-class mainstream politics. The skilled may have a life style which is closer to the middle class' but a political rhetoric which resonates with working-class images. This paradox encourages pragmatic politics (cf. Hamilton, 1964:53-57; Michels, 1959:292-294). Perhaps, in advanced industrial societies, skilled workers may become a special-interest group within the union and vacillate in their ideological loyalties.

11

The Social Construction of Anomie

Most workers in the four nations were aware of the major problems confronting their unions, cities, and nations, but they were not active in organizations attacking these problems. The reasons for noninvolvement may vary from one society to another. In some societies, the problems may be so enormous that workers feel that nothing can be done about them, while in others they may see problems as being adequately managed. Even within a society, workers may disagree whether society is integrated or disorganized and anomic. Sociologists have observed that people who are involved in nonlocal social systems tend to see society as integrated, while the uninvolved feel anomic. This chapter examines how the system involvements of workers are related to their beliefs about societal anomie.

No sociologist has, to my knowledge, measured societal anomie directly,[1] probably because the concept is too ambiguous and our measurement technology too primitive. Yet sociologists have neither clarified the concept nor abandoned attempts to measure it. They have instead psychologized it and devised scales to measure how people *feel* about vague statements about the world (cf. Srole, 1956; Dean, 1961). Importantly, these scale results have been interpreted not as cognitive beliefs about societal norms, but as evidence of the respondent's own deviance (Merton, 1938; Cohen, 1965), poverty and ignorance (Meier and Bell, 1959), or personal alienation (Roberts and Rokeach, 1956).[2] Although some students (Nettler, 1965) have suggested that anomie scales really measure personal despair, and others think that the concept is useless (Feuer, 1963; Lee, 1972), most sociolo-

[1] Angell (1951) measured social (moral) integration, the opposite of anomie, in cities.

[2] Purists have insisted that, since anomie is a societal condition, individuals cannot be anomic. The personal response to living in an anomic society and the feeling of being unattached to society are said to be forms of alienation (Seeman, 1959).

gists have not abandoned the concept.[3] On the contrary, they have spawned more concepts of self-feelings (meaningless, powerlessness, political inefficacy, self-estrangement) and have tried to identify the characteristics of people with these feelings.

The search for social correlates of anomie has produced data which corroborate the existence of social stratification rather than societal anomie. Thus, people who score high on anomie scales tend to be poor, uneducated, and nonparticipators in organizations controlled by people in higher strata; the low scorers tend to be middle-class, educated, and organizationally active (cf. Mizruchi, 1960; McClosky and Schaar, 1965; McDill and Ridley, 1962). But students of anomie do not like this simple stratification interpretation of their findings. Marxists insist that scholars must also pay attention to the industrial and capitalistic contradictions which objectify and alienate the workers (Israel, 1971:50-62), and Durkheimians feel that societal normlessness is more complex than simply a matter of who has what. Unfortunately, neither Marxists nor Durkheimians have demonstrated how their central concepts should be studied empirically.

Since measurement of societal anomie is not currently possible and since a psychological interpretation is less than satisfactory sociologically, a third approach may prove more fruitful: interpreting responses to anomie scales as cognitive data.[4] The task then becomes one of explaining how position in the social structure affects beliefs about specific societal strains, rather than identifying who feels deviant, rejected, or anomic. The cognitive approach makes no assumptions about the relationship between the extent of societal anomie observed and personal feelings about it. The received view that anomie scales measure personal confusion should be rejected in favor of the view that they measure estimates of societal normlessness. I hypothesize that the system involvements of strata are structured differently in various societies and that beliefs concerning societal anomie reflect how people in various strata see the effects of societal arrangements upon them. This chapter examines how stratal position (skill level) and social system involvements of workers affect

[3] Although Feuer (1963) and Lee (1972) objected to the concept of alienation, their criticisms apply also to anomie as a form of alienation.

[4] In such an event, anomie scale items should clearly describe normless societal situations rather than feelings about conditions in society.

their belief about the extent of anomie in their societies. The findings are interpreted in the context of specific problems which confront different strata in each society.

Theoretical Perspective

Any study of anomie should consider both work and its societal setting. Although the informants of this study all lived in the urban-industrial sectors of their societies, the technology and social organization of the sectors differed. The more industrial the society, the more complex were the technology of automobile production and the more complex were factory, community, and societal organizations. Technology and affluence tended to go together; workers in the most developed nations were most affluent and those in the least developed nations were poorest. The most industrialized nations had had most experience with manufacturing, most time to accommodate their institutions to a complex economy, and most opportunity to mediate conflicting societal norms. The poorer nations were struggling to accommodate their institutions to new industrial demands and to devise norms which would harmonize with the changing situation (Moore, 1967:75-87). In short, contrary to conventional wisdom, the urban-industrial sector of the least industrial society probably had the least integrated social organizations and norms (Kerr *et al.*, 1960). I arbitrarily assumed that the United States had the most integrated (least anomic) urban-industrial sector and India the least integrated, with Italy and Argentina in between.

On the basis of such structural differences in the urban-industrial sectors, certain hypotheses may be derived about the extent of societal anomie workers observe. First, since highly industrial nations have more successfully solved problems of institutional integration, their workers observe more societal order than workers in less industrial societies. Second, since the technology and social organization of newly industrializing societies are simpler than in older societies, workers in new industrial societies exhibit more consensus on the extent of anomie. Third, the extent of anomie which workers see in their own societies reflects differences in stratal exposure to conflicting norms and societal disorganization. The anomie which workers in different strata perceive varies with the societies because the strata are exposed to different economic, political, and social forces in each society.

The picture in the United States is unambiguous: since the labor force is highly urbanized and well educated, the unskilled recognize the contradiction between societal mobility norms and their own immobility. They tend to be absorbed into local social systems such as the family, neighborhood, and church, and feel they have no place in the wider society. Understandably, they observe more societal anomie than the skilled, who are better paid, better educated, and more involved in nonlocal systems such as the union, political party, community, and national associations. They understand larger social systems and their place in them and thus perceive the society as more orderly than do the less skilled.

This congruence of income, education, skill level, participation, optimism, and political conservatism with low anomie has been observed in highly industrial societies (Inkeles, 1966; Almond and Verba, 1965; Curtis, 1971). However, the situation may well be reversed in less developed societies which have changing stratification systems (Hoselitz, 1964). As an elite minority of the urban manual labor force, factory workers are disproportionately exposed to new technologies, new organizations, and rapid social changes. How much anomie they observe depends upon many factors. Unskilled and semiskilled workers who have migrated from rural areas have achieved considerable mobility merely by getting factory jobs. Their lives may be so absorbed by local systems (work group, family, and neighborhood) that they remain oblivious of wider societal problems. Thus they may consider both their lives and society as quite orderly.

In less industrial societies, skilled workers may perceive more anomie than the less skilled. Being better educated and participating more in nonlocal systems, the skilled become better informed about the real and pressing national problems. Even though they themselves are the least deprived sector of the working class, they feel that they deserve continuous upward mobility because their skills are so scarce (Germani, 1966:373). If blocked, they may conclude that the system cannot or will not solve its problems, that the system lacks norms, order, and justice. Traditional elites aggravate this situation by refusing to recognize the status claims of industrial workers (Simpson, 1970; Lopreato and Hazelrigg, 1970). Finally, skilled workers are frustrated by their inability to link the work group, union, party, community, and other associations to increase their political influence. Altogether, these conditions heighten their perception of societal normlessness (Smelser and Lipset, 1966:23-29).

I was not satisfied with most anomie scales because their items describe feelings of pessimism or frustrated ambitions rather than societal normlessness.[5] I tried to devise statements describing the absence of norms, contradictory norms, or norms which changed too rapidly.[6] After pre-testing a number of items, I settled on the following seven:

1. In the everyday problems of life, it is easy to know which is the right path to choose.
2. It is hard to rear children nowadays because what is right today is wrong tomorrow.
3. It seems that nobody agrees on what is right and wrong because everyone is following his own ideas.
4. It is easy to find agreement on what is morally right.
5. There are so many organizations with different goals that it is impossible to trust any of them.
6. The world is changing so fast that it is difficult to be sure that we are making the right decisions in the problems we daily face.
7. The man with morals and scruples is better able to get ahead in this world than the immoral and unscrupulous person.

In each country, different sets of five items formed Guttman scales which had coefficients of reproducibility of 0.87 or above.[7]

Level of Industrialization, Skill, and Anomie

Three hypotheses were suggested above: (a) the less industrial the nation, the greater the consensus on the extent of societal

[5] An exception is Olsen's study (1969), which distinguished two views of anomic political culture held by two strata in a community. See also Taviss's classification of self-society anomic relationships (1969).

[6] Using this standard, Dean's (1961) scale is superior to Srole's (1956) and McClosky and Schaar's (1965) is superior to Dean's. Seeman's scale (1959) emphasizes socially unapproved means to achieve given goals and not normlessness.

[7] OLDS, 0.88; FIAT, 0.87; IKA, 0.89; PAL, 0.87. Statements 1 and 4 were dropped for OLDS, 4 and 5 for FIAT, 5 and 6 for IKA, and 1 and 7 for PAL. Ranges for agreement to the statements were: OLDS, 31-79%; FIAT, 47-79%; IKA, 46-87%; PAL, 55-75%. The standard deviations were: OLDS, 1.33; FIAT, 1.22; IKA, 1.50; PAL, 1.70.

anomie, (b) the more industrial the nation, the lower the level of anomie, and (c) the less industrial the country, the more anomie skilled workers observe in their societies.

The expectation that workers in the less industrial societies would display highest agreement on the extent of societal anomie was partly supported. With one exception (Argentina), the range in agreement to scale items increased with industrialization: India, 20 percent; Italy, 32 percent; Argentina, 41 percent, and the United States, 48 percent.[8] However, the standard deviation, as a measure of consensus, showed higher agreement in the more industrial societies: FIAT employees had the lowest standard deviations, followed by OLDS, IKA, and PAL. Apparently, consensus around the central tendency increased with industrialization, but so did the range of scores. This simultaneous homogenization and differentiation of beliefs about societal anomie which accompany industrialization will be explained later.

The hypothesis that workers in the most developed society would observe least societal anomie was also partly supported. Except for Argentina (whose workers had the lowest median scores), the more industrial the nation, the less anomie workers observed (see Table 11.1). Some features of the Argentinian situ-

TABLE 11.1 DISTRIBUTION OF ANOMIE SCORES (PERCENTS)

Guttman Scale Scores		*OLDS*	*FIAT*	*IKA*	*PAL*
Low	0	7	2	28	13
	1	21	8	22	6
	2	24	18	16	5
	3	24	31	21	16
	4	18	28	10	28
High	5	6	13	4	32
Total		100	100	100	100
Median		2.4	3.2	1.5	3.9
Mean		2.4	3.1	1.7	3.4
N		(247)	(306)	(275)	(262)

[8] Carr (1971) suggests that in the application of the Srole scale, the acquiescence effect increases with poverty. I believe that my items avoid most of the difficulties of the Srole scale. Other interview evidence revealed a willingness of workers, especially in Argentina and India, to disagree with the status quo and received views.

ation may explain the low scores. Córdoba was the fastest-growing industrial city in the country, and IKA's employees had the best and highest-paid jobs in the city. The union was exceptionally effective both in the plant and in national politics. Except for the United States, the Argentinians were the best educated, best housed, and best fed workers of the four nations. These conditions support de Imaz's (1970:271-272) observation that Argentina has attributes of modernism "in excess of" its economic development.

Finally, the third hypothesis was partly supported: the lower the nation's industrialization, the more societal anomie skilled workers observed. In all factories, the skilled received about 50 percent higher wages than the lowest-paid workers, yet only in the United States did the skilled see their society as more highly integrated (see Table 11.2). Surprisingly, despite the alleged rigidity of the Italian stratification system, no differences in anomie appeared by skill level. In IKA, the skilled polarized slightly while the semiskilled tended to concentrate in the middle range. In India, both unskilled and skilled observed higher anomie than did the semiskilled. Thus, in three of the four countries, the position of the skilled differed from that in the United States pattern. Clearly, different factors were associated with the normlessness that people in various strata observed in their societies.

Social System Involvements and Anomie

Two explanations for anomie are dominant in the social sciences: the Marxist, which emphasizes economic deprivation and the dehumanizing aspects of factory work, and the Durkheimian, which stresses the absence of overlapping social bonds.[9] According to either explanation: (a) more upwardly mobile than nonmobile workers should believe that society is orderly and not anomic; (b) employees with routine jobs should observe more anomie than those with nonroutine jobs; (c) isolated workers should see more anomie than workers with many social contacts on and off the job; (d) people most identified with the union and most active politically should see society as most integrated; and (e) employees with many organizational ties in the community

[9] Strictly speaking, Marx was not concerned with anomie, but with alienation. See note 2 above.

TABLE 11.2 ANOMIE ACCORDING TO SKILL LEVEL (PERCENTS)

	Anomie Scores				
	Low (0,1)	*Middle (2,3)*	*High (4,5)*	*Total*	*(N)*
OLDS[a]					
Unskilled	26	46	28	100	(72)
Semiskilled	22	50	28	100	(151)
Skilled	40	52	9	100	(83)
$p = <.01$					
$\bar{C} = .230$					
FIAT					
Unskilled	12	48	40	100	(94)
Semiskilled	9	49	42	100	(156)
Skilled	11	48	41	100	(56)
$p = >.10$					
IKA[a]					
Unskilled	50	36	14	100	(134)
Semiskilled	49	42	9	100	(111)
Skilled	55	23	22	100	(70)
$p = <.05$					
$\bar{C} = .180$					
PAL					
Unskilled	22	14	64	100	(95)
Semiskilled	19	28	53	100	(102)
Skilled	14	21	65	100	(65)
$p = .10$					
$\bar{C} = .244$					

[a] These data are for the analytic samples, which contain additional skilled workers.

and nation should see the society as more integrated than those with solely local ties (e.g., family and neighborhood).

Socioeconomic Background, Skill, and Anomie

Data in Table 11.3 demonstrate that in all countries workers at different skill levels had different social origins, educational opportunities, and occupational and economic experiences. However, these variables were not related to anomie, nor were related variables such as community of origin, father's occupation, education, occupational experience, duration of unemployment, and career or generational mobility. Only in PAL was there a tendency for respondents to see more societal anomie if most of their employment had been in manufacturing and if they had been upwardly mobile both generationally and in their own careers. The major finding is that social origins, opportunities, and occupational and economic experiences were related to skill level but not to beliefs concerning societal anomie.

TABLE 11.3 COEFFICIENTS OF CONTINGENCY BETWEEN SOCIOECONOMIC AND OCCUPATIONAL-BACKGROUND VARIABLES FOR SKILL AND ANOMIE

	OLDS		*FIAT*		*IKA*		*PAL*	
Variables	*Skill*	*Anomie*	*Skill*	*Anomie*	*Skill*	*Anomie*	*Skill*	*Anomie*
Father's occupation	—	—	.21[a]	—	.26	—	—	—
Community of socialization	.24	—	.42	—	—	—	.35	—
Education	—	—	.35	—	.51	—	.54	—
Sector work experience	.22[a]	—	.36	—	.27	—	.49	.24[a]
Unemployment	—	—	.24	—	.28	—	—	—
Occupational mobility	.43	—	—	—	.39	—	.48	.26
Generational occupational mobility	.62	—	.48	—	.65	—	.38	.25[a]

[a] The probabilities of the χ^2s upon which the coefficients of contingency are based are between .06 and .10. Unless otherwise indicated, the probabilities for all the associations in the following tables are at or below the .05 level.

Work Environment, Skill, and Anomie

Although the job monotony and social isolation associated with automobile production allegedly produce dissatisfied and alienated workers, little support for these ideas was found in any of the factories. Job satisfaction was relatively high everywhere; most did not find their jobs monotonous, nor did they want to change their job routines. Skill level was positively associated with job satisfaction everywhere and, in three of the plants, it was associated with positive job evaluation. In OLDS and IKA, skill level was positively associated with intimacy of contacts among workers (see Table 11.4).

While assemblers allegedly perform the most alienating operations in auto production, in no factory did assemblers see more societal anomie than other workers. In all factories, employees in test, inspection, and repair and machine operators in three fac-

TABLE 11.4 COEFFICIENTS OF CONTINGENCY BETWEEN JOB SITUATION VARIABLES AND SKILL AND ANOMIE

	OLDS		FIAT		IKA		PAL	
Variables	Skill	Anomie	Skill	Anomie	Skill	Anomie	Skill	Anomie
Job satisfaction	.299	—	.278	—	.198	—	.243	.195
Desire to change work routines	.335	—	—	—	.318	—	—	.276
Occupational evaluation	.594	—	—	—	.314	—	.238[a]	—
Work attachment	—	—	—	—	—	—	—	—
Number of good friends at work	—	—	—	—	.179[a]	—	—	—
Ratio of interaction opportunity realized on the job	.296	—	.486	.345	—	—	—	—
Index of quality of interaction with work mates	.380	—	—	—	.295	—	—	—
Index of quality of outplant interaction with work mates	.222	—	—	—	—	—	—	—

[a] The probabilities of the χ^2s upon which coefficients are based is between .06 and .10.

tories saw more anomie than assemblers. Moreover, except for PAL, anomie was not associated with the worker's job satisfaction, nor was it related to the desire to change work routines, occupational evaluation, work attachment (would quit working if guaranteed present income), job aspirations, and the type of employment preferred (manufacturing, service, or farming).

Further, beliefs about societal anomie were not associated with amount or type of social contacts on the job, i.e., how much respondents liked their work mates, the number of good friends at the factory, the amount of interaction on the job, and the extent to which workers took advantage of interactional opportunities. Finally, no relationship was found between beliefs concerning anomie and amount of work-mate interaction off the job. One exception to these findings should be noted. In PAL, those who wanted to change their work routines felt that society was more anomic than those who did not, and more workers who saw a moderate amount of anomie tended to be more satisfied with their jobs than those who saw either less or more. This anomalous situation will be explained later.

The conclusions to which these findings lead are clear: even though job satisfaction and work-group cohesion were related to skill level, especially in OLDS and IKA, neither work monotony nor work isolation was associated with the anomie workers observed in society. These results fail to support both Marxist and Durkheimian notions: impoverished background, industrial socialization, occupational immobility, unemployment, monotonous work, and social isolation on and off the job do not affect the extent of anomie believed to exist in society. These data provide indirect evidence that, in responding to the anomie scale, workers were not giving vent to their personal frustrations, but giving their thoughts about what was going on in the nation.

Union and Politics, Skill, and Anomie

Following Durkheim, some scholars have demonstrated that workers who are most attached to their unions are also most involved in community and national affairs (Dean, 1954; Form and Dansereau, 1957; Tannenbaum and Kahn, 1958; Spinrad, 1960), and that those most involved in political affairs exhibit least anomie (Milbrath, 1965:78-79). In this study, an extensive analysis has shown that skilled workers in all factories except PAL were more involved in union affairs than other workers were. However, only in OLDS was high union participation associated with beliefs of high societal integration. Nowhere were high interest in union affairs and high political activity associated with these beliefs (see Table 11.5). Only PAL employees who wanted militant unionism tended to see their society as highly anomic and only political radicals in IKA saw their society as more anomic than other workers. In sum, while position in the skill hierarchy was associated with union participation, only in the United States was participation associated with the extent of anomie observed, and only in IKA was political ideology related to the observed anomie. Here again, the data suggest that beliefs about societal anomie are not dependent on the extent of the workers' union involvement or on their political ideology.

Nonwork Social Systems, Skill, and Anomie

Sociologists have long observed that people of low estate are locked into local social systems (family and neighborhood) and people of high socioeconomic rank are more involved in nonlocal organizations of the community and nation. Sociologists have also concluded that people involved in nonlocal organizations see

TABLE 11.5 COEFFICIENTS OF CONTINGENCY BETWEEN INDICATORS OF UNION AND POLITICAL INVOLVEMENT BY SKILL AND ANOMIE

Variables	OLDS		FIAT		IKA		PAL	
	Skill	*Anomie*	*Skill*	*Anomie*	*Skill*	*Anomie*	*Skill*	*Anomie*
Interest in unions	—	—	—	—	.200	—	—	—
Participation in union affairs	.360[a]	.314	.329	—	.280	—	—	—
Most important union functions	—	—	.273	—	.274	—	—	.283
Amount of union militancy desired	(—).271	—	—	—	—	—	—	.267[a]
Evaluation of union performance	—	(—).268[a]	—	—	.257	(—).257	—	—
Amount of political activity	.538	—	N.A.	N.A.	N.A.	N.A.	—	—
Radical-conservative ideology	—	—	—	—	.259	.272	—	—
Social-class identification	(—).226	—	N.A.	N.A.	.214	—	.374	—
Tenacity of beliefs	—	.254	—	—	—	—	—	—

[a] The probabilities of the χ^2s upon which coefficients are based are between .06 and .11.

their society as orderly and integrated. Thirty indicators were devised to measure involvement in the family, neighborhood, community, and national systems, and composite indices were constructed for participation in each system. Except for FIAT, nowhere was skill associated with family or neighborhood involvement. Only in OLDS was skill related to community involvement, but in all nations except PAL, skill was related to the index of national involvement (see Table 11.6). Only in the United States were a majority of involvement indicators in any major system (the community) associated with the anomie observed. In short, the anomie which workers perceived in society was not related to their involvement in the four major systems: family, neighborhood, community, and nation.

To summarize: the amount of anomie which workers observed in their societies was not associated with their occupational mobility, economic experiences, monotony of work, social cohesion with work mates on or off the job, the extent of involvement in union or political affairs, or involvement in the family, neighborhood, community, or nation. These findings strongly suggest that beliefs concerning societal anomie are held independent of personal experiences and social involvements, and that such beliefs are to be taken at face value, as cognitive estimates of societal anomie.

Latent Patterns of System Involvement

Since weak but consistent trends in the data may be hidden by arbitrary statistical requirements, the data were reexamined for hints which might support Marxist or Durkheimian theory. Those in the bottom third of the distributions were defined as observing societal integration, those in the top third, as observing anomie. Different patterns appeared for each nation.

United States

Employees who saw society as integrated were city dwellers who had higher than average interaction with their neighbors, claimed to have more friends, were more active in the union and other organizations, and participated in more organizations with their work mates. Those who saw society as anomic tended to be young semiskilled workers who preferred the factory to employment in other sectors. They wanted the union to stress social sol-

TABLE 11.6 COEFFICIENTS OF CONTINGENCY BETWEEN SELECTED INDICATORS OF FAMILY, NEIGHBORHOOD, COMMUNITY, AND NATIONAL INVOLVEMENT BY SKILL AND ANOMIE

	OLDS		FIAT		IKA		PAL	
Variables	*Skill*	*Anomie*	*Skill*	*Anomie*	*Skill*	*Anomie*	*Skill*	*Anomie*
Visits relatives on vacations	.365	—	—	—	.267[a]	—	N.A.	—
Relatives live in neighborhood	—	—	—	—	—	—	.281	—
Index of familism	—	—	.258	—	—	—	—	—
Number of friends in neighborhood	—	—	—	—	.267	—	—	—
Exchanges visits in neighborhood	—	(—).27	.204[a]	—	.249	.27	—	—
Discusses neighborhood problems	—	—	—	—	—	—	—	—
Church attendance	.252[a]	—	—	—	—	—	—	—
Index of neighborhood involvement	—	—	—	—	—	—	—	—
Interest in local news	—	—	—	—	—	—	—	—
Names community problems	—	.39	.300[a]	—	—	—	.306[a]	—
Organizational memberships	.362	(—).26	—	(—).30	.255	—	.249[a]	—
Fellow workers member of organization	.440	(—).30	—	(—).30	—	.26	—	—
Index of community involvement	.354	(—).23[a]	—	—	—	—	—	—
Names national problems	.348	—	.291	—	—	—	—	—
Interest in national news	—	—	—	—	—	—	—	—
Index of national involvement	.311	—	.286	—	.259	—	—	—

[a] The probabilities of the χ^2s upon which coefficients are based are between .06 and .10.

idarity but they were not interested in politics. In their free time they engaged in spectator activities (watching movies, television, and sports events) and participated little in organizational activities. In short, the pattern for OLDS roughly conformed to traditional findings.

Italy

Those who saw Italian society as relatively integrated tended to be highly educated men whose fathers were skilled or white-collar workers. They tended to be oriented to the neighborhood, to avoid personal ties with fellow workers but to participate in religious organizations with their neighbors who were also workmates. They frequented the local bar, belonged to a local religious organization, and had friends who resided in the neighborhood, but they saw no neighborhood problems. As sympathizers of conservative unions, they rejected militant tactics and favored union cooperation with management as the way to improve their economic benefits. Although they were liberal but not dogmatic in their politics, they thought that the major national problems were moral. In brief, these workers had tightly organized their lives on a folk-urban neighborhood basis. They had accommodated to the dominant authority structure but nonetheless saw society as fairly integrated. No pattern was discerned for workers who saw society as highly anomic (cf. Bonazzi, 1964:67-101).

Argentina

Employees who saw society as integrated displayed no distinctive pattern. Those who viewed it as anomic tended to be skilled and pro-union. However, they wanted the union to press for economic rather than political goals. Although they claimed to have many friends, they did not visit them, nor did they attend organizational meetings with friends or fellow workers. Although slightly more involved in neighborhood and community affairs than the average employee, they were also more disenchanted with the national political scene, which they saw as chaotic. In short, although their roots in all social systems were shallow, they were deeper than those of workers who saw society as relatively integrated.

India

Employees who saw Indian society as relatively integrated tended to be rural migrants who had experienced more than average upward mobility. Somewhat more involved than others in the union, they wanted it to be politically active but were themselves inactive and conservative. In contrast, respondents who saw society as moderately anomic were highly educated semiskilled workers who had worked only in factories all their lives. Though they claimed to have many friends, they spent more time alone than did the average worker. Their neighborhood ties were few; they had little interest in neighborhood, union, or community affairs; but they had greater interest in national politics. Radical in political ideology, they wanted militant unions though they did not participate in any arena. Workers with high anomie scores, as in FIAT, displayed no distinctive pattern.

The search for trends resulted in four patterns, but only OLDS followed the "classic" pattern of workers with high involvement in nonlocal systems to perceive societal orderliness. Italians who believed their society was integrated exhibited a folk-urban pattern of neighborhood involvement and avoided commitments to work and nonwork organizations. They accepted managerial control of their work world, preferred nonmilitant unions and liberal political parties. Argentinians who saw society as integrated were indistinguishable from others, but observers of high anomie had developed more than average neighborhood and community ties, were attached to the union and its laboristic politics, and had become disenchanted with national political affairs. Finally, India had the most complex pattern. Observers of high societal integration were upwardly mobile rural migrants interested only in the union and its economic performance. Respondents who saw society as anomic displayed no distinctive characteristics. Those who saw moderate anomie approximated the IKA pattern for observers of high anomie: high education, low occupational mobility, low involvement in all systems except the national, and radical political ideology. The four different national patterns suggests that workers were responding to distinctly different societal conditions.

System Correlates of the Skill-Anomie Relationship

Since sociological theory points to the critical importance of system involvement to anomie, I decided to examine how skill and anomie are related when system involvement variables are controlled. In examining the results, we should keep in mind the original relationship between skill and anomie with the variables under consideration. I shall summarize them before presenting data with the system involvements controlled.

Social and Economic Background

I have demonstrated that skill was strongly associated with social origin and occupational mobility in all countries, but anomie was not. The correlations between skill level and anomie beliefs for social and economic background variables are found in Table 11.7. In all nations except India, skilled workers reared

TABLE 11.7 COEFFICIENTS OF CONTINGENCY BETWEEN SKILL LEVEL AND ANOMIE FOR SELECTED BACKGROUND CHARACTERISTICS OF AUTOWORKERS

Background Characteristics	*OLDS*	*FIAT*	*IKA*	*PAL*
Community of socialization (metropolitan)	(–).371	(–).401	(–).354	–
Education (OLDS, low; IKA and PAL, high)	.398	–	(–).411	(–).377
Occupational mobility (OLDS and IKA, up; PAL, none)	(–).407	–	(–).310	.392
Generational mobility (above father)	(–).420	–	(–).425	–
Duration of unemployment (none)	(–).370	–	(–).293	.469
Father's occupation (factory)	(–).411	–	–	–
Major sector of previous employment (manufacturing)	(–).294	–	–	–

in metropolitan areas saw society as more integrated than did other workers. For FIAT, no other background conditions affected the skill-anomie relationship, but in OLDS and IKA, the traditional relationship appeared: the higher the skill, the more societal integration was observed by workers who had experienced upward career and generational mobility but no unemployment. In PAL, the higher the skill, the more anomie was perceived by workers who had experienced neither unemployment nor upward occupational mobility. In sum, the skill-anomie relationship persisted in OLDS, emerged in IKA, did not appear in FIAT, and appeared inconsistently in PAL.

Job Satisfaction and Worker Relations

While the skilled in all societies were the most satisfied with their jobs, only in OLDS and IKA did they have relatively more contact with their work mates. Nowhere were anomie beliefs associated with job satisfaction and the quantity or quality of contacts with work mates. However, when job satisfaction and interaction with work mates were controlled for OLDS, the higher the skill, the more societal integration was observed by those most satisfied with their jobs and in most contact with their work mates on and off the job. The opposite situation existed in IKA: for those with most job satisfaction and most work-mate contacts, the higher their skill, the more anomie they observed. A similar but less consistent pattern was found in PAL, and the pattern was weakly visible in FIAT (see Table 11.8). This anomaly of work

TABLE 11.8 COEFFICIENTS OF CONTINGENCY BETWEEN SKILL AND ANOMIE WITH JOB SITUATION VARIABLES CONTROLLED

Item	*OLDS*	*FIAT*	*IKA*	*PAL*
Job satisfaction (satisfied)	(−).311	–	.313	.312
Desire for change in work routines (no; PAL, yes)	(−).512	–	.417[a]	.361
Occupational self-rating (good)	(−).397	–	–	–
Work attachment (would continue working even if guaranteed pay)	(−).349	–	(−).367	–
Number of good friends at work (high)	(−).288	–	–	.209[a]
Contacts with fellow workers (satisfied; FIAT, not satisfied)	(−).589	.549[a]	.233	.238
Ratio of work interaction with work mates (high)	(−).300	–	.381	–
Quality of interaction with work mates (high)	(−).296	–	.600[a]	(−).380[a]
Quantity of outplant interaction with work mates (high)	(−).366	–	.301	–

[a] The probabilities of the χ^2s upon which the contingency coefficients are based are between .06 and .10.

satisfaction and worker solidarity to be associated with high skill and high anomie will be examined later.

Union Attachment and Politics

In all nations except India, the skilled participated the most in union affairs, but skill differences in political participation were small. Although some scholars (Seeman, 1971) claim that people who are uninvolved in organizations and politics feel powerless and anomic, no supporting evidence was found in this study.[10] In OLDS, skill level and anomie beliefs were inversely related (high skill and low anomie) for every measure of union and political involvement: interest in unionism, positive evaluation of unions, high participation in union affairs, and high participation in political activity. Indicators of union involvement in FIAT followed the OLDS pattern, in which the skilled tended to see society as integrated. In IKA and PAL, the situation was partly reversed; for those highly involved in the union, the higher their skill, the more societal anomie they saw. Finally, among the political conservatives and neutrals in OLDS and FIAT, the higher their skill the less anomie they observed. In IKA and PAL, the higher their skill, the more anomie political radicals saw in their societies (see Table 11.9). In short, skill and beliefs about anomie were negatively associated with union and political involvement in OLDS; a similar but weaker pattern emerged in FIAT, and the opposite pattern appeared in IKA and PAL.

TABLE 11.9 COEFFICIENTS OF CONTINGENCY BETWEEN SKILL AND ANOMIE FOR CONDITIONS ASSOCIATED WITH UNION AND POLITICAL INVOLVEMENT

Variables	*OLDS*	*FIAT*	*IKA*	*PAL*
Interest in unions (OLDS, high; IKA and PAL, none)	(–).505[a]	–	.369[a]	.436
Participation in union affairs (high)	(–).407	–	.335	.423
Most important union function[b]	(–).372	–	.332	–
Amount of union militancy desired[c]	(–).450	(–).373	(–).403	–
Evaluation of union performance (good; PAL, poor)	(–).312[a]	(–).560	(–).264	.596
Social class identification[d]	(–).356	N.A.	–	.424
Amount of political activity (low)	(–).450[a]	N.A.	N.A.	–
Radical-conservative ideology[e]	(–).323	(–).444[a]	.455[a]	.406[a]
Tenacity of beliefs (low)	(–).368	–	–	–

[a] The probabilities of the χ^2s upon which the contingency coefficients are based are between .06 and .10.

[b] OLDS, cooperate with management; IKA, promote worker solidarity.

[c] Alternatives: militant, determined, bargain, cooperate. OLDS, determined; FIAT, bargain.

[d] OLDS, working class; PAL, lower middle.

[e] OLDS, neutral; FIAT, conservative; IKA, liberal and radical; PAL, radical.

[10] For OLDS, scores on a scale of powerlessness were not associated with anomie scores.

Nonwork Social Ties

Only in OLDS was skill level positively related to community and national involvement, and the absence of community involvement related to perceptions of high societal anomie. High family and neighborhood ties were found among skilled workers in FIAT and moderately high ones in IKA, but no pattern was discernible for PAL. The absence of organizational ties was associated with beliefs in high anomie only in OLDS and FIAT.

When involvement in nonwork systems was controlled, different patterns appeared in each country. In OLDS, the skilled (compared to the less skilled) observed more societal integration when their family and neighborhood ties were minimal and when their community and national ties were well developed (see Table 11.10). In FIAT, no patterns appeared, but in IKA statistically significant relationships appeared, especially in the family and neighborhood areas, and they were generally positive: when family involvement was low, the higher the skill, the higher the belief in societal anomie. The same relationship between skill and anomie persisted when neighborhood or national involvement was high. Only three associations reached levels of statistical significance in PAL, but whatever the strength of involvements in the neighborhood, community, or nation, the skilled consistently saw society as more anomic than did the less skilled (see Table 11.10).

Skill and Anomie: A Summary

The relationships between skill and anomie for all the social systems are summarized in Table 11.11. In OLDS, the two variables were related in the expected direction in every system from the family to the nation: the skilled saw society as more integrated when involvement with the family and neighborhood was low and when involvement with the job, work mates, union, community, and nation was high. In FIAT, skill was not associated with observation of societal anomie, whatever the system involvements of workers. This is puzzling, given the importance of status distinctions in Italian society (LaPalombara, 1965:309-316). Apparently, the Italians sensed considerable societal anomie regardless of their system involvements (cf. Almond and Verba, 1965: 308-310). Italian factory workers appeared to constitute a homogeneous proletariat because even the skilled and organizationally involved saw as much anomie as did unskilled nonparticipants.

TABLE 11.10 COEFFICIENTS OF CONTINGENCY BETWEEN SKILL AND ANOMIE FOR INDICATORS OF NONWORK SOCIAL SYSTEMS

Arenas of Involvement	*OLDS*	*FIAT*	*IKA*	*PAL*
Family arena				
Visits relatives on vacation	–	–	(–).495	N.A.
Wife works (OLDS, yes; IKA, no)	(–).347	–	.347	–
Relatives in neighborhood (none)	(–).336	–	.336[a]	–
Index of familism (low)	(–).324[a]	–	.315	–
Neighborhood arena				
Number of friends in neighborhood (few)	(–).314	–	.340[a]	–
Exchanges visits with friends (moderate)	.324	–	.604	–
Names neighborhood problems	–	–	–	–
Church attendance (never)	(–).487	–	.335[a]	–
Index of neighborhood involvement (OLDS, medium; IKA and PAL, high)	(–).364	–	.395[a]	.393[a]
Community arena				
Interested in local news (high)	(–).482	–	–	–
Names community problems (institutional)	(–).415	–	(–).308	–
Member of organizations (none)	.298	–	–	–
Work mate is member of common organization	(–).488	–	–	–
Index of community involvement	(–).361	–	–	.335[a]
National arena				
Names national problems	–	–	–	.382
Interest in national news (high)	(–).565	–	–	–
Index of national involvement (high)	(–).450[a]	–	.340[a]	–

[a] The level of significance of the χ^2s upon which the coefficients are based is between .06 and .10.

All workers seemed to define present economic conditions as tolerable but the future as uncertain.

The Argentine situation contrasted sharply with the Italian. Despite the governmental instability in Italy, the economic "miracle" was unfolding; productivity and living standards were rising, and unemployment was decreasing. In contrast, inflation, unemployment, strikes, and political crises were racking Argen-

TABLE 11.11 COEFFICIENTS OF CONTINGENCY BETWEEN SKILL AND ANOMIE FOR INDICATORS OF INVOLVEMENT IN VARIOUS SOCIAL SYSTEMS

Indicators	*OLDS*	*FIAT*	*IKA*	*PAL*
Familism index (low)	(–).346	–	.472[a]	–
Job satisfaction (high)	(–).311	–	.314	.314[a]
Quality of contacts with work mates (high)	(–).297	–	.614	–
Quantity of contacts with work mates outside plant (high)	(–).377	–	–	–
Union participation (PAL, medium; others, high)	(–).413	–	.335	.423[a]
Neighborhood involvement index (OLDS, medium; IKA and PAL, high)	(–).364	–	.396[a]	.393[a]
Community involvement index (OLDS, high; PAL, low)	(–).361	–	–	.334[a]
National involvement index (high)	(–).448	–	.340[a]	–
Political ideology[b]	(–).323	(–).444	.455[a]	.406[a]
Overall social interaction	(–).375	–	–	–

[a] Probabilities of the χ^2s upon which coefficients are based are between .06 and .10.

[b] OLDS, neutral; FIAT, conservative; IKA, liberal and radical; PAL, radical.

tina. Yet IKA's employees saw less anomie in their society than FIAT's did in theirs, and the skilled in IKA who were most satisfied with their jobs, most involved with their union, fellow workers, neighborhood, and nation saw their society as more anomic than the less skilled. Why should this be?

The unskilled in IKA, as in FIAT, appeared to be relatively content with their economic lot and unconcerned with what was happening in the city and nation. While they considered the government inept, they had well-paid and relatively stable jobs and thus did not see the urgent need for social and political changes. On the other hand, a minority of the skilled were highly literate and well informed about the deteriorating position of labor in the industrial sector since the Peron era (Soares, 1966:190-197). Their despair did not result from a sense of detachment from society, from ignorance about how their society was functioning, or from a poverty of ideas on how to improve the situation. They argued cogently on what was wrong with the economy and polity. In so doing, they were not as resigned to the status quo and working-class status as were the Americans and the Italians (Fillol, 1961: 90-92).

A situation not unlike Argentina's appeared in India, but it was not as fully developed—as is indicated by the fewer correlations and their lower levels of statistical significance (see Table 11.11). Again, the unskilled saw society as relatively integrated; with well-paid jobs and secure employment, they saw order. However, a minority of well educated semiskilled and skilled workers saw a great deal of societal normlessness. Not only were they the most informed about the massive problems facing their country, but they also felt that the system had failed to provide them with jobs appropriate to their status and training. Their response was to criticize all organizations as political or corrupt, and to refuse to join any of them. Unlike workers in other countries, the best educated participated least in social organizations (cf. Sheth, 1968: 110-112).

Political Ideology Among the Observers of High Anomie

Since people who see their society as anomic may become either apolitical or radical, I decided to examine their political ideology in detail. Ideology may provide a clue to the strategy people desire to reduce anomie: conservatives typically want to reduce governmental activities so that a natural harmonious order reappears; radicals want government to rearrange institutions to reduce the unnatural anomie. Exploring the relationship between anomie and ideology may be instructive because both variables, as defined, deal with cognitive views of the world.

Data in Table 11.12 show no differences in the skill composition of OLDS conservatives and liberals who saw high societal anomie. In FIAT, however, the percentage of unskilled and semiskilled workers who saw society as highly anomic increased with conservatism, and the proportion of the skilled increased toward the radical pole. A similar situation appeared in IKA, except that the skilled were ideologically polarized. PAL workers displayed an irregular pattern: the percentage of unskilled remained about the same among all ideologies, the semiskilled were underrepresented among the radicals, while the skilled, as in IKA, split between both conservatives and radicals.[11] Except for the United

[11] The company classified many long-tenured low-skilled workers as skilled. The use of this classification (rather than the researcher's) revealed,

TABLE 11.12 POLITICAL IDEOLOGY OF WORKERS WITH HIGH ANOMIE SCORES BY SKILL LEVEL (PERCENTS)

	Political Ideology					
Company	*Apathetic*	*Very Conservative*	*Conservative*	*Neutral*	*Liberal*	*Radical*
OLDS:						
Unskilled	—	—	29	28	39	—
Semiskilled	—	—	57	65	50	—
Skilled	—	—	14	6	11	—
Total	—	—	100(14)	99(32)	100(18)	—
FIAT						
Unskilled	50	—	43	—	21	22
Semiskilled	50	—	46	—	59	44
Skilled	0	—	11	—	20	33
Total	100(18)	—	100(28)	—	100(53)	99(27)
IKA						
Unskilled	—	—	67	47	33	20
Semiskilled	—	—	22	20	11	40
Skilled	—	—	11	33	57	40
Total	—	—	100(9)	100(15)	101(9)	100(10)
PAL						
Unskilled	—	35	36	—	42	43
Semiskilled	—	46	29	—	36	26
Skilled	—	19	35	—	21	31
Total	—	100(43)	100(45)	—	99(33)	100(35)

Note: Ns shown in parentheses.

States, a substantial proportion of skilled workers who saw their society as anomic was concentrated among the radicals, while the less skilled tended to embrace a more conservative ideology (cf. Lipset, 1960:240-242). These results encourage speculation that beliefs about anomie and ideology are associated with specific societal conditions.

among those who believed that Indian society was highly anomic, a greater proportion of skilled conservative workers and a smaller proportion of radicals. The company's classification also augmented the proportion of radical semiskilled workers but reduced the proportion of unskilled workers.

Discussion

American sociologists have interpreted responses to anomie scales as personal feelings about the world rather than as beliefs concerning the normative integration of society. With respect to the United States, this personality-test approach labels individuals as anomic, psychologically ineffective rejects. A more sympathetic view sees anomic people as lower-class victims of a stratified order (McClosky and Schaar, 1965). Both views are sometimes applied uncritically to findings in other societies. Thus, the studies of Simpson (1970) and Soares (1966) both see anomic Latin American workers as victims of frustrated mobility aspirations and fail to observe Durkheim's suggestion that workers' social-system involvements be examined within the context of institutional relationships.

I have suggested that responses to anomie scales be interpreted as cognitive estimates of societal normlessness. Support for this position was provided by data which showed that different strata in the four nations see their societies as anomic. In the United States, where a stable stratification system allocates rewards among manual workers, the skilled receive the highest rewards, participate in organizations which affect their welfare, and therefore see society as integrated. The unskilled, who observe that society is not organized to serve their interests, do not participate in its organizations and believe that societal norms are vacuous slogans (Huber and Form, 1973).

But this pattern may not describe the situation in less industrial societies, where unskilled factory workers from rural areas experience upward mobility, feel relatively secure, adhere to the norms of their communities of origin, remain unaware of societal problems, and therefore see the society as orderly (Allardt, 1965). On the other hand, highly educated and economically privileged skilled workers who identify with the middle class may be well informed about institutional crises and see their society as anomic. Such a situation can occur in several societal settings. Moderately industrialized societies, like Italy, which have strong class cleavages and a radical political tradition, may socialize some workers who have extensive organizational networks ranging from the work-place to the union and party (cf. Portes, 1971), to view their society as anomic. In societies with high levels of education, moderate industrialization, and unstable politics, such as in Argentina, people in the same stratum may differ in their estimates of national anomie (de Imaz, 1970:217-220). Highly educated middle-class skilled workers may believe society is anomic because they see that political and economic institutions fail to realize the norms of providing the educated with continuing economic mobility. They embrace a radical ideology in the hope of changing both the society and their position in it. The less industrial the nation, the more readily this kind of situation can arise.

The temptation to generate a theory which explains everything is great. But sociologists do not know whether industrialization creates a predictable sequence of societal problems which dispose different strata to sense varying amounts of anomie. Apparently a simple technological explanation will not do. For example, FIAT workers lived in a technological environment similar to OLDS's, yet the Italians responded differently to it. The responses of OLDS and IKA workers were in some respects similar, even though their technological environments differed considerably. Yet in each society, the findings were intelligible in the context of local history, the system of stratification, and the problems confronting workers in the factory, union, community, and nation. The appropriate type of interpretation is strongly suggested in the literature on political movements (Janowitz, 1970: Section II; Lipset, 1970:227-265).

Interview data in this study provide clues on what workers respond to in arriving at an estimate of societal anomie. It seems to be the extent to which they see themselves integrated into the political systems of the union and nation and the extent to which they see these two systems as mutually responsive. Those who are unaware of this issue are politically apathetic and tend to see society as orderly. If the issue is recognized and the union and nation are seen as reciprocally responsive, society is seen as orderly; if the issue is recognized, but union and nation are not reciprocally responsive, society is seen as anomic. Different strata in the society may interpret the national scene differently.

In the United States, the political systems of the union and nation are integrated and stabilized. Since skilled workers have the most access to these systems, they see the society as integrated. In the three other poorer countries, unskilled industrial workers are relatively happy with their economic lot and are insufficiently integrated into the union or political system of the nation to perceive problems of societal integration. In Italy, the political fragmentation of unions and parties is so complex that most workers give up trying to make sense of the pattern; hence their conviction that the society is anomic (LaPalombara, 1965). A few skilled workers in the radical unions and parties have embraced an ideology which defines the class-union-nation linkages as even more anomic.

During the Peron era, Argentine labor unions were integrated into the national polity and workers who were active in the union probably considered the society as integrated. At the time of the study, unions still had power at the plant level, but their political influence was officially contained at the national level. Skilled workers, who were the most active in the union, were the most aware of the economic consequence of political exclusion; hence their belief in high societal anomie. If unions are admitted into a polity which they can influence, the skilled, like their American counterparts, may become more conservative and see the society as orderly.

India represents an extreme case. Whatever the union's potential strength, workers saw India's problems as beyond early resolution, hence the widespread recognition of anomie. Moreover, the union's weakness in the factory and in politics was recognized especially by the educated semiskilled and skilled workers who also felt that their mobility was blocked by the particularism of both management and the union. The fact that neither personal nor national problems were likely to be resolved in the near futured led some educated Indians to opt for radical political solutions. Yet they were unwilling to join radical organizations, perhaps because of their strong middle-class identification.

The explanation for societal anomie beliefs offered here has the virtues of being sociological, of giving workers credit for being able to explain how they fit into the larger society, and of being researchable. The sociological task is to study how well beliefs concerning anomie correspond to data obtained from structural studies of system relationships and societal norms. For this task, survey research merely opens a window.

12

Technology, Participation, and Stratification

The master trend of the past two centuries has been the gradual industrialization of the world. Social scientists interested in the impact of technological change on class organization have often assumed that machines would first dehumanize and alienate workers, but eventually that machines would homogenize, unify, and politicize them. This study has examined one organizational aspect of this complicated problem: whether industrial workers associate more with one another as societies industrialize, whether they become more involved in organizations, and whether these personal and organizational involvements create the basis for a solidary working-class movement. Theoretically, the solidarity of the working class may increase or decrease over time, or one stratum within the class may become more cohesive and organizationally influential while another remains unorganized. Such stratification would diminish the chances for a successful working-class movement.

Two hypotheses dealing with this problem were examined. The industrialism hypothesis, made famous by Alex Inkeles, predicts that wherever the factory appears, workers quickly develop a common mentality and similar organizational involvements. The factory is the school for making people modern in their thoughts and behavior. While a critical mass may be necessary to launch a class movement, widespread industrialization is not a prerequisite. Development theorists such as Wilbert Moore and Neil Smelser see industrialization as a process whereby workers gradually accommodate to factory life, shed their family and local ties, and become involved with the world beyond the factory. The opportune time to launch a working-class movement is as soon as workers see themselves as a class, as soon as they create organizations to serve their purposes, and as soon as they link these organizations to political action. If leaders do not act in time, workers may get involved in social systems which are not class-linked and lose their fervor for class politics.

Technology is a central variable in both the industrialism and developmental hypotheses, in two senses. First, factory technology stimulates the formation of work groups which can be linked to external organizations such as the union and political party. Second, the growth of technology creates the organizational environment which links the factory and union to national political systems. The industrialism and developmental hypotheses are complementary rather than contradictory. Each powerfully explains different phenomena at different stages. The industrialism hypothesis explains the social psychological readiness of workers to be involved in social movements, while the developmental hypothesis explains the organizational prerequisites for such movements.

I have tried to simplify the study of this problem by selecting four countries which vary in industrialization (India, Argentina, Italy, and the United States) and, within these countries, by selecting only one industry—automobile manufacturing. Then I have stratified the samples into skilled, semiskilled, and unskilled workers to ascertain how different technological exposures affect their behavior in the work group, union, party, neighborhood, community, and nation.

Before reviewing the results, I acknowledge that factors other than technology, such as national traditions, institutions, values, ideologies, and economic problems may affect class solidarity. I tried to take them into account in interpreting the data and tentatively concluded that an increasingly complex technology tends to stratify rather than homogenize the working class. However, since conflict between classes may be more politically consequential than the internal stratification of one class, the research findings must be interpreted cautiously.

In this study, the stratal position of autoworkers varies by country: the more industrially advanced the nation, the lower they are in the stratification system. Thus, in the United States, where all workers are educationally qualified for low-skilled industrial jobs, automobile factories tend to hire the least educated low-class workers. In India, where general educational levels are low and where good industrial jobs are scarce, autoworkers are recruited from the educated middle classes.

The stratification of the working class, as manifested by skill levels, also varies with national industrialization. Since skilled workers operate the most complex machines in the most industrial countries, and since unskilled operations remain relatively unchanged in all countries, skill differences among workers increase with industrialization. The skill hierarchy becomes both more differentiated and more elongated. Contrary to expectations, skilled workers have higher prestige in the most industrialized nations. Although all factory workers constitute a relatively privileged stratum in less developed countries, in advanced societies the mass of factory workers are thought not to have distinctive occupations and are therefore accorded little prestige, while skilled workers are honored. Yet the skilled constitute an envied elite in all countries. They not only have received more formal education and vocational training, they have experienced less unemployment and more upward mobility. Everywhere the skilled have started life with more opportunities than the less skilled and have retained their advantages in the factories and elsewhere.

A precondition for forming solidary work groups is the opportunity to interact on and off the job. I assumed that conditions which permitted more interaction also encouraged worker cohesion and solidarity, and that conditions which reduced interaction promoted cleavages. While factory technology is the most important variable influencing worker interaction, other factors may be important. For example, differences in socialization may aggravate cleavages among workers. Thus, rural- and urban-born workers may be physically segregated in the factory because the rural may hold the least desirable jobs. Moreover, they are said to be clannish and reluctant to join working-class organizations. If this is true, rural-urban cleavages should be most profound in the least industrial countries, where rural-urban cultures differ the most. Predictably, except for the United States, more workers of rural origin were unskilled and more of the urban-born were skilled. But for all other factors which might affect working-class solidarity, rural-urban differences were minimal. Both groups adapted well to their jobs and factory life; both interacted with about equal frequency with their work mates on and off the job, they had similar interests and attitudes toward the union, and both participated at approximately equal rates in union, neighborhood, community, and national affairs. Although overall system involvements of the rural group were slightly lower than for the urban, neither saw society as more anomic than the other. In short, community background was not associated with factors related to working-class solidarity and rural-urban cleavages were not more profound in the less industrial countries.

Since workers interact more on the job than away from it, the extent of work-group cohesion probably reflects the interaction opportunities jobs provide. Skilled workers have most opportunity to interact because they have most control over their complex machines, while the unskilled are bound to automatic machines and have less freedom to move about. With advancing technology, the machines which the skilled operate get more complex, giving them more autonomy and opportunity to make contact with work mates. But at the same time, as technology advances, machines become larger, dispersing workers spatially and reducing their opportunity to interact. The overall effect of advancing technology should be to reduce job interaction in general and to increase the differences in interaction rates among the skill levels.

Data from this study verified these generalizations. The lower the factory's technological complexity, the higher was the density of its work stations, and the more workers took advantage of their interactional opportunities. The more complex the technologies, the lower was the density of work stations and worker interaction rates. Moreover, the quality of intimacy of interaction was weakly associated with work-place density, being highest in the factories with the simplest technologies. But the amount and quality of interaction was greater for skilled than unskilled workers everywhere. As predicted, because the skilled operated the most complex machines in the most industrial countries, the quantity and quality of their contacts were highest in those nations. Finally, although worker contacts outside the factory were highest at OLDS, they were low everywhere. Overall, these findings suggest that the simpler technologies permitted more interaction than did the advanced technologies, especially for unskilled and semiskilled workers. Insofar as working-class solidarity is stimulated by work contacts, it should decrease with advancing industrialization because cleavages along skill lines are greater in the most industrialized societies.

Regardless of workers' opportunities for communication inside and outside the factory, a working-class movement must be born of a sense of grievance over poor working conditions and inequitable distribution of rewards. There is no research consensus on whether workers feel more oppressed by factory work at early or late stages of industrialization. Some scholars claim that when workers are first exposed to factory labor, they rebel against the strict scheduling of work, the artificial physical environment, and the monotonous routines. Other scholars, Marxists especially, believe that technology becomes increasingly oppressive as it develops and that workers in the most industrial countries rebel most against work schedules, factory environment, and job monotony.

Data from this study failed to support either position consistently. In the newer industrial societies, employees accepted work fatalistically, as part of their lives. They did not expect it to be satisfying or fulfilling, but the desire for satisfying work was strongly developed and generally satisfied among respondents in the mature industrial societies. Most workers found factory employment more attractive than farm or service employment, but factory was most appealing in Italy and Argentina, where the manufacturing sector was growing most rapidly. In all countries, the highly skilled found the factory most appealing. Finally, both direct and indirect methods of studying jobs satisfaction showed that workers in all countries were satisfied with most aspects of their jobs. Predictably, the higher their skill and the more control they had over their work, the more satisfied they were. Importantly, in the less industrial societies, unskilled workers were relatively more satisfied; in the more industrial countries, the skilled were relatively more satisfied. Again, dissimilarities in satisfaction were greater in the more advanced countries.

Contrary to the view that the unskilled see their jobs solely in economic and instrumental terms, the majority in all countries wanted interesting labor. Yet assemblers were not more concerned about job monotony than others, nor were they seeking different jobs. For most employees, the important issue was not monotony but the quality of working conditions and how management was utilizing their skills. Finally, the opportunity for job interaction was more important for unskilled than for skilled workers. The overall findings are clear: technology did not appear to be more repressive in either early or mature industrial societies; in all societies, employees accepted work as important in their lives; they preferred the factory to other types of employment; and though satisfied with their jobs, they hoped and expected that their children would be upwardly mobile. The sense of grievance needed for an aggressive working-class movement was not widespread. These findings, once thought to apply uniquely to Americans, were found to be universal.

Working-class movements do not arise spontaneously; they are consciously organized by labor unions and political parties working together. Since the union is closer to the worker than the party is, in the first stage of political mobilization the union must convince workers that it must engage in political action. This is not an easy thing to do. Theorists disagree on the optimum conditions for socializing workers for political unionism (Portes, 1974). Some (Kuczynski, 1967) hold that industrialization quickly radicalizes workers to accept political unionism, while others (Bauman, 1972) emphasize that a full-blown industrial society is necessary for successful political unionism. But both views agree that workers most subjected to the machine (e.g., assemblers, machine operators) are most alienated, and, when aroused, most susceptible to radical appeals. On the basis of this factor alone, political unionism should be most acceptable to semiskilled workers in highly industrialized countries.

But union ideologies vary in their stress on political mobilization. The Italian unions were the most dedicated to political activism, followed by the Indian, Argentinian, and American. However, the data showed that neither the extent of industry in the nation nor the union's devotion to politics affected the members' views on the primary goals of the unions. Unions were everywhere accepted as necessary and legitimate, but nowhere did workers feel that a primary union goal should be to change the economic and social institutions of the nation. Instead, they stressed that unions should pursue economic gains and the improvement of working conditions. The less industrial the country, the more workers pressed unions for economic improvements; the more industrial, the more they wanted better working conditions. These findings support a developmental theory of technological effects: where technology is most widespread and complicated, workers are more concerned about its effects on working conditions than its potential to increase their incomes.

In all countries, though a small minority of workers endorsed political unionism, a large minority actively opposed any political activity on the part of their officers. Even political radicals opposed political unionism. Beliefs in radical politics, political unionism, and militant unionism were not correlated in a consistent pattern in the four countries. The most militant unionist were often politically conservative, and radicals were often politically apathetic, but radicalism and political unionism were most highly associated in the rapidly industrializing nations, Italy and Argentina.

Contrary to expectation, assembly and other semiskilled workers did not support political unionism more than other workers did, but they tended to be more militant unionists. In OLDS and especially in FIAT, they pressed unions hard to improve working conditions. In all factories, the skilled were more concerned with protecting their economic gains through orderly bargaining rather than through militant tactics. Although the skilled were the most active union members everywhere, they dominated union affairs more in the mature industrial societies. With the possible exception of IKA, skilled workers were both the most conservative and the most involved in national politics. They were also the best informed politically and the most active politicians in the factories, dominating communication in a conservative and self-interested way. Even in IKA, where they were more radical than other employees, the skilled wanted union officers to emphasize economic benefits and working conditions rather than the pursuit of political goals. The cleavage between officers and members over political unionism was most severe in the more industrial countries, where the skilled were foremost in their opposition and where the semiskilled were most concerned about working conditions. Thus, in accord with the development hypothesis, the spread of technology seemed to stimulate union political cleavages.

While unions and parties play a central role in coordinating working-class movements, workers will not become involved in them unless they understand how factory, union, community, and national problems are linked. The ability to understand this linkage occurs when workers escape the particularistic bonds of family and neighborhood, and when they become involved in union, community, and national affairs. If officers prematurely push workers to political action, they risk being attacked as self-serving politicians; if they urge political action after workers have already become involved in nonlocal systems, the time to launch a social movement may have passed, because workers will have accepted a system of pluralistic politics. The industrialism hypothesis holds that the introduction of the factory system quickly frees workers from local bonds and involves them in a class movement to change the political economy of the nation. The development hypothesis holds that it takes time to involve workers in organizations which extend from the factory to the nation, and that genuine working-class movements naturally develop in mature industrial societies (Smelser, 1959).

I attempted to measure the extent that the family and neighborhood absorbed the free time of workers as well as the extent that they became exposed to the problems of the broader community. Surprisingly, even in the least industrial nation, the family and neighborhood did not monopolize the workers' free time. Everywhere they had developed substantial contacts with their work mates and others in the community, especially during the weekends and holidays. Everywhere they were informed about the major problems of the union, neighborhood, city, and nation, and a minority were heavily involved in organizational activities. These findings tend to support the industrialism hypothesis of early readiness for social movements.

Yet, even in the most industrialized society, no more than one-quarter of the workers were highly involved in systems extending from the factory to the nation. These were the activists who linked local systems and external systems to energize social movements. The data here tend to support the development hypothesis, that the effective linking of organizations appears with mature industrialism. Thus, the Americans who were most active in the work group were also involved in the union, community, and national organizations, while the Indians who were active in one system were not involved in others. The American activists shared the political ideology and class identities of their fellow workers. The Italian, Argentinian, and Indian activists were all more radical than their work mates. Yet in Italy and Argentina workers who were active in both union and national affairs had few if any ties in the work group or the community. In India, workers with most social contacts in various social systems were more radical than their fellow workers, but they avoided union and political responsibilities and identified more with the middle than the working class.

In summary, though activists in the United States had linked various systems successfully, they were conservative, had been absorbed into the pluralistic political system. In Italy and Argentina, activists who were highly interested in union and national politics lacked a political base among the workers. In India, activists talked radical politics, but they did nothing. Perhaps the time to launch a working-class movement had passed in the United States, was still possible in Italy and Argentina, and had not yet appeared in India.

The political involvement of industrial workers is made easier when they are homogeneous in beliefs and social system involvements—i.e., when they have the attributes of a social class. An important question is whether the cleavage between the skilled and other workers is aggravated or moderated by the industrialization process. The evidence in this study was that the skilled were better able than others to forge solidary work-groups and to carry them over into union and community. Moreover, they were less absorbed in family and neighborhood networks and more involved in community and national affairs. The more industrial the society, the clearer these trends appeared to be. Apparently, skill differences in social system involvements from the family to nation are smallest at the earliest stage of industrialization. What binds industrial workers at that point is their common identity as factory workers, but what separates them later is their skill differences and system involvements. In short, the working class appears to become more stratified with increasing industrialization.

But the more important question may be whether the strata become more or less alike in their ideologies. With one exception, the more industrial the country, the more conservative were the skilled workers. Moreover, their political aggressiveness and union militancy were not affected by their politics but by local economic and political conditions. Clearly, while wage differences among the skill levels remained constant, in the more industrial societies, the skilled were more concerned with preserving their advantages. Perhaps, with increasing industrialization, the skilled became less excited about a working-class movement and more concerned about their own special interests. Whether they supported the middle or the working class depended on how local conditions affected them.

The final condition which affects workers' dispositions to join class movements is their estimate of the amount of disorder in society. Those who see it as disorderly or anomic may either become apathetic or join a movement of the left or right to bring about order. Those who perceive society as orderly usually want to keep it the way it is. Thus political beliefs and beliefs about the extent of societal anomie tend to be related. Whether industrialization increases societal anomie and whether workers reflect the change are disputed social science issues which involve political commitments. A reasonable position would be Durkheim's view that organizationally involved groups are likely to see society as integrated. An important exception would be political extremists, who may belong to organizations dedicated to overthrow an anomic social order. The organizationally uninvolved may either be apathetic about the issue or see society as anomic.

According to development hypothesis, workers in mature industrial nations probably see more societal integration because their societies have had more time to achieve normative and organizational integration, while workers in less developed societies see more anomie because their societies are struggling with massive and sometimes unmanageable problems. On the basis of previous findings, we should expect skilled workers to see society as more integrated because they participate more in organizations, especially in advanced industrial societies. Since the stratification of the working class may increase with industrialization, disagreement on the amount of societal anomie should also increase with industrialization.

As expected, the data showed that the less industrial the society, the more anomie workers observed. Argentina was the exception because its workers perceived the least anomie. As hypothesized, the more industrial the society, the more was the dissensus on the extent of anomie. However, the expectation that the skilled would perceive the least anomie was supported only in the United States. In the less developed societies, the unskilled tended to see society as relatively orderly, while many skilled and educated workers saw it as more anomic. The explanation for this finding calls for specific stratal knowledge of worker experiences in each nation. The skilled and the educated in less developed societies are better informed about national problems than the less skilled. Sometimes the status claims of the skilled are denied by traditional elites, with the result that the skilled are excluded from civic and other organizations. Consequently, some workers embrace a radical ideology which premises a state of societal anomie.

A general proposition may be offered to explain whether skilled workers perceive societal anomie. If they believe that the union is effectively tied into national political decision-making, and if they are involved in politics, they see society as orderly. If they believe labor to be politically ineffective and are themselves uninvolved in unions, they see more societal anomie. For this reason, the chances of a successful working-class movement are greater in less developed societies, where skilled workers are aware of national problems, the role unions can play in politics, and their common interests with less skilled workers. The success of a movement depends in great measure upon the union officials' ability to involve skilled workers in the union and in politics. Failure to do both tends to be the rule.

Discussion

In the analysis of the structural conditions underlying a working-class social movement, technology played a conspicuous but not a determinative role. One persistent problem was my hesitation to generalize on the basis of only four cases. The two middle cases of Italy and Argentina were particularly troublesome either because they were too similar for our purposes or because, as middle cases, they varied more than cases at the extremes. While Italy was slightly more industrial than Argentina, the latter's population was better educated and more urban. Thus, in some areas, the responses of Argentinian workers were more like the Americans'; in other areas they were closer to the Italians'. Support for the development hypothesis might be stronger by treating the two nations as a single middle case. In one respect, the United States, Italy, and Argentina were quite close to each other when compared to India. Cases like Korea, Taiwan, and Greece would be useful additions as intermediate cases of industrialization. On the other hand, the study would have profited by adding mature industrial countries such as Great Britain, Sweden, and Germany, which have moderately successful working-class labor movements. The addition of eastern European countries would not be so useful because they lack an independent labor movement.

A more severe problem than insufficient number of cases is the absence of a theory which considers how industrial development, stratification, and social movements are related. The industrialism and development frames of reference constituted crude guideposts of what might be expected, but I had to improvise a workable theory as I went along. The industrialism hypothesis, first specified by Inkeles and later elaborated by Kerr and his associates, has the advantage of being clear and straightforward. I applied it to areas the authors had not considered, and found it moderately powerful in predicting how workers would respond to the technology of the work place. Despite variations in the cultures of nations, and variations in technology, workers responded similarly to work-values, factory environment, their specific jobs, and their immediate work groups. But the industrialism hypothesis was less powerful in predicting system involvements beyond the work place. Union involvement was predicted better than was involvement in other social systems, because the union is more closely linked to the job. Thus skilled workers who had the most work-autonomy formed the most solidary work-groups, and they used this solidarity to control the local union. The more technically advanced the society, other things being equal, the more this seemed to be the case. One feature of less developed societies is that their governments try to involve workers in national affairs and unions serve as intermediaries between the worker, the party, and the nation. Except for the United States, involvement in the union induced involvement in national affairs. But the nature of the union-party-nation organizational ties differed in each country.

In India, both the union and the nation were too distant from workers to command their involvement. They could not see how either could solve the problems confronting the nation. Even the most radical failed to join organizations because they saw society as too anomic. In Italy, workers apparently understood the union-party-nation nexus, but they saw both unions and parties as too competitive to be able to exert enough influence in national affairs to help them. If unions united, they could bring order into the political demands of workers and create an effective working-class movement because workers understood the need for one. But the ideological fragmentation of unions into communist, socialist, liberal, Catholic, and pro-business factions prevented working-class solidarity more than did the special interests of different skill categories. Obviously conservative interests profited from the ideological fragmentation of the unions.

In Argentina a strong working-class social movement was constrained more by governmental policy than by cleavages among workers. Skilled workers, especially, remembered the first Peron regime in which labor had had more influence, and they realized that its economic and political influence had deteriorated. Workers appeared ready for political mobilization but, in the face of government resistance, many opted for the best economic improvements they could get under the existing system. In a democratic pluralistic polity, Argentinian workers might easily have formed a class party or become a labor faction within a dominant party because they were not so torn by ideological factions as Italian workers were. Of the four nations, the Argentinians were in the best position to launch a strong working-class movement.

In the American case, the labor movement had been absorbed into the mainstream of national politics. While many workers realized that the union represented their interests in the Democratic Party, they did not see these interests as representing a class ideology. Though the majority preferred Democratic electoral victories, they did not believe that such victories would bring them economic gains. These had to be won by collective bargaining, but here workers' interests were split. The UAW, responding to the majority of unskilled and semiskilled workers, had obtained across-the-board wage improvements, and decreased the wage differential between the skilled and the less skilled. The skilled, wanting to maintain their traditional advantage over the less skilled, demanded veto power over contracts negotiated with management. Many skilled workers thought that they could bargain better as members of craft unions than as members of industrial unions. In the political arena, these workers were suspicious of the recommendations of UAW leadership. In short, the workers in the most industrialized nation were most divided in their economic, social, and organizational interests.

Conclusions

The purpose of this research was not to formulate a theory of working-class social movements, but to examine how the technology of society and the factory could influence workers' involvements in social systems which tie into such movements. The theory proposed in Chapter 1 was generally supported. Factory technology differentiates interaction opportunities of workers, forging groups of varying solidarity along skill lines. With increasing industrialization, these groups become more differentiated in their union and organizational involvements, possibly increasing the stratification of the working class.

The working class may become even more stratified in the future. Why should this be so? First, technological innovation will continue to change the interactional opportunities of occupational groups at different rates. Automation has continued to disperse workers spatially, reducing the interaction of some while increasing the consultative activities of others (Smith, 1968). The total effect has been to differentiate interactional opportunities even more than mechanical production. Second, as technology advances, industrial workers become a smaller proportion of the labor force. Eventually, industrial unions may represent a small special-interest group and not a large homogeneous working class. As the proportion of organized labor declines, it loses its underdog status.

More white-collar employees, now laboring under conditions similar to those of manual workers, may become organized into a salariat similar to the traditional proletariat. Yet the growth of white-collar unions in the United States has not matched the proportionate decline of manual workers in the labor force. Moreover, the distance in status between white-collar and manual workers shows little evidence of decreasing. Closer to traditional and influential elites, white-collar workers jealously guard their status advantage. Even if all non-professional white-collar workers unionized, the status distance between them and industrial workers would be greater than the current distance between unskilled and skilled industrial workers. The conflicts among semiprofessional, technical, clerical, and manual workers do not dissolve when they become unionized (Windmiller, 1974, 1975).

Finally, unionized workers, both manual and white-collar, will probably remain more concerned about their own security than about the well-being of the poor and unorganized manual workers in the service sector. At one time, labor spoke for all manual workers on the assumption that they had common interests. But while craft and industrial unions improved conditions for their members, the conditions of poor workers in the service sector remained unchanged. The urban poor now seem as much a feature of mature industrial societies as they are of developing countries.

Economists have recently recognized that the manual sector

is stratified into a dual labor market. The primary market is made up of large quasi-monopolistic unionized firms, such as auto, steel, and aluminum manufacturers. Because these firms control prices and routinely pass wage increases on to consumers, wage levels are relatively high. Government workers, who are now organizing, constitute almost a fifth of the labor force of advanced industrial societies. Like workers in the primary market, they have begun to carve out a secure niche for themselves. The secondary labor market, composed of economically marginal small businesses, hires uneducated workers distinguished by race, ethnicity, and sex. These enterprises are rarely unionized and, if unionized, pay wages not much above those of the unorganized sector. Low wages, unemployment, irregular employment, and business failure characterize the secondary labor market. So while labor and management in the primary sector and labor in the governmental sector increase their well-being, low wages and economic insecurity are a constant feature of the secondary market.

Perhaps the most important cleavage of the future working-class will be between those who have relatively secure incomes and those who do not. With threatening unemployment, workers become more concerned with the stability rather than the level of income. Those most likely to find it will be in the primary labor market, and they will struggle to keep whatever advantages they have. The more successful industrial, white-collar, and government workers become in achieving economic stability, the more exposed will be the position of unorganized poor, increasing the stratification of the working class.

The situation in industrializing underdeveloped societies will probably deterioriate. Already industrial workers constitute an elite of the working class, and recent rural migrants form a miserable subproletariat who live in the slums surrounding large cities. If such nations import a more advanced technology, an even higher stratum of technical workers will form above the industrial workers, resulting in a three-stratum class of technical, industrial, and marginal workers.

In both developed and underdeveloped societies, the ideology of economic equality has taken hold. The unorganized poor demand to share the fruits of technological progress. Organized labor probably will not fight to distribute some of its gains to the poor. In all societies, the working class may become more stratified if wages are allocated on the basis of organizational strength rather than equity principles. Only government can control the tendency of unchecked technological change to produce more economic inequality. But who will put pressure on government to regulate technology and redistribute economic rewards? Social movements with this goal must attract people from all classes who believe that people rather than technology direct social change.

Appendices

Appendix A

Evolution of the Study

I have long been interested in automobile workers. At the World Congress of Sociology in Stresa, Italy, in 1958, I discussed my interest with Professor Filippo Barbano of the University of Turin. He encouraged me to carry on my research in Turin. I also met Paolo Ammassari, who had studied industrial sociology at the University of Florence, and who would collaborate in future research. I planned to replicate at FIAT studies I had done at OLDS. The research question would be, "Is Italian social structure sufficiently different from that of the United States to affect how autoworkers respond to industrial and urban life?" My intellectual bias led me to believe that the Italians would be much like the Americans.

Differences between Italy and the United States seemed important enough to justify a study. Although the automobile industry in Italy was as old as the American, only during the recent Italian economic miracle had production approached American proportions. Italian industry still responded to the traditional crafts, while mass production dominated American industry. Italian industry was absorbing a large number of southern migrants who had had little or no industrial experience. This phenomenon was about over in the United States. Italy presumably had a more rigid stratification system than did the United States, and Italian industrial workers had a proletarian tradition of political unionism in contrast to the American tradition of pragmatic bread-and-butter unionism. The research would explore whether differences in extent of industrialism, craft versus mass-production emphasis, rural versus urban backgrounds of workers, rigid versus open stratification systems, political versus pragmatic unionism affect the ways in which workers respond to their industrial environment.

Paolo Ammassari came to the United States and we designed the study. I learned to speak Italian fairly well before going to Italy in June of 1961. During the following year in Turin, we worked out the details of the study. After some difficulties (see

Appendix B), we succeeded in interviewing a sample of FIAT workers. Refusal rate was about 7 percent. Professor Ammassari, responsible for directing the field work, later wrote a monograph dealing exclusively with the FIAT workers (Ammassari, 1964; 1969). After returning to the United States, I decided to replicate the Italian study. Steven E. Deutsch directed the field work at OLDS during the 1962-63 academic year. As in Italy, 7 percent of the workers contacted either refused to be interviewed or could not be located.

The more familiar I became with the Italian and the American data, the more I was convinced that the study should be expanded to include factories in countries at lower levels of industrialization. Turin was a much more industrial and cosmopolitan city than I had expected. Southern Italians were not being employed in large numbers in FIAT, which preferred the better-educated migrants from northern Italy. I had been thwarted in attempts to gather detailed information about the political and union identifications of FIAT workers. I sought opportunities to study automobile factories in Latin America and Asia, and in 1965 I received funds to enable me to expand the research.

Professor Delbert C. Miller had been in Córdoba, Argentina, and had made good contacts with the sociologists at the university there. He told me about IKA, which had started automobile production in the city in 1955. The city and factory seemed to qualify as a research site. Córdoba was in the interior of the country, sufficiently far from Buenos Aires to have an independent industrial development. Richard E. Gale went to Córdoba to direct the field work. He had good command of Spanish, but was fortunate to employ Virginia Robledo, who helped train and supervise the interviewers. I spent a month in Córdoba supervising the translation of the interviews and making contacts with university, union, and company officials. Despite some difficulties in getting cooperation from union and company officials, we finally obtained good-quality interviews, with a refusal rate of less than 4 percent. Gale's (1969) independent analysis of the IKA materials emphasized problems of social stratification and labor mobility.

In July, 1965, Baldev R. Sharma went to India to locate a research site. After visiting several factories, he selected PAL in Bombay. Immediately he confronted the difficult problem of translating the interviews into the mother-tongues of the workers: Marathi, Hindi, Urdu, Gujarati, and Punjabi. Fortunately he was fluent in English, Hindi, and Urdu. Interviewers who spoke other languages were trained for their tasks. The most important problem they faced was to make sure that the workers understood the questions. On the average, three hours were needed to complete each interview, about twice the time required in the other countries. On the whole, satisfactory interviewers were obtained, with a refusal rate of only 3 percent. Field work was finished in May, 1966. Sharma (1969, 1974) extended the analysis to include materials on absenteeism, turnover, and labor commitment.

Before going into the field, the researchers learned as much as they could about the community and the factory to be studied. Local contacts were made, but in no case did we receive formal permission to conduct the studies. All of us were able to converse in the required languages prior to departure for the field. Upon arrival, three to five months were spent in learning the history of the community, the factory, the union, and local industrial relations. Newspapers and periodicals were reviewed, interviews were held with local union, management, and government officials. We talked to industrial workers, visited their homes and neighborhoods, and, when possible, attended union meetings. In two instances (FIAT and IKA), sociologists at local universities provided the initial contacts with union and management officials, but most contacts were made without sponsorship. After we felt that we understood the local scene sufficiently, we asked permission of union and management to carry out the study. In all cases it was obtained after some difficulty (see Appendix B). We also received permission to inspect the factory, to familiarize ourselves with the physical and technological environments of the departments to be studied.

After this period of orientation, we translated the interview and adapted it to the particular features of the plant, union, community, and society. In translating the interview, where possible, we involved local sociologists who knew English and sociology students from manual-worker backgrounds. After revisions and pre-tests, we interviewed workers who were not in departments selected for study. Letters describing the research and informing workers of management and union cooperation were mailed to workers prior to making contact for interviews. Except for India, we interviewed workers in their own homes. Difficulty in locating addresses in Bombay, the lack of privacy in the homes, and management's willingness to provide ideal interviewing facilities persuaded Dr. Sharma to conduct the interviews in an office removed from the work-site. The foreman released respondents during work-time. While we do not know how much this procedure affected responses, workers expressed considerable hostility toward both management and the union during the interviews, suggesting that they were not intimidated. The fact that the interview did not focus on collective bargaining or on the evaluation of management probably made it relatively nonthreatening.

TABLE A.1 OCCUPATION OF FATHERS (PERCENTS)

	OLDS	*FIAT*	*IKA*	*PAL*
Farmers	35	34	21	35
Unskilled	14	18	10	9
Service	1	13	12	6
Semiskilled	18	6	20	4
Skilled	21	18	7	9
Clerical	1	6	4	6
Proprietors	3	4	22	19
Managers and professionals	6	–	4	12
Total	99	99	100	100
No. of cases	(240)	(297)	(270)	(257)
Mean rank	3.2	2.9	4.1	4.0

TABLE A.2 COMPANY AND RESEARCHER SKILL CLASSIFICATIONS FOR MAHARASHTRA CASTES AND NON-HINDUS

	Researcher's Classification			*Company Classification*			
	Un	*Ss*	*Sk*	*Un*	*Ss*	*Sk*	*Totals*
Brahman	20	27	43	—	67	33	100
Kshatriya	19	43	38	15	36	49	100
Village servants	45	25	30	25	10	65	100
Other Hindus	44	34	22	12	47	42	100
Non-Hindus	44	45	11	7	50	43	100

Researcher's Classification: $\chi^2 = 26.811$, d.f. $= 8$, $p = <.01$, $\overline{C} = .3716$.
Company Classification: $\chi^2 = 22.207$, d.f. $= 8$, $p = <.01$, $\overline{C} = .3451$.

Appendix B

Field Problems in Comparative Research: The Politics of Distrust*

Why should anyone trust a snooping sociologist? In my first comparative study along the United States-Mexican border, I became aware that informants in the two countries responded differently to the same research strategy. In subsequent comparative studies, I observed even greater differences. This chapter reviews my encounters with distrustful respondents in this study. I hope that my observations will stimulate others to publish their field protocols, because their analyses might suggest ways to manage problems of distrust more effectively.

I began to think systematically about field problems of comparative research shortly after returning from Italy in 1963. My colleagues had asked me to report my findings in a departmental colloquium. Since the data had not been processed, I decided to describe the vexing field problems I had encountered. In Italy, Paolo Ammassari and I constantly tried to persuade reluctant organizations to cooperate in our study. My interest in industrial sociology led me to analyze these efforts within a collective-bargaining frame of reference, and I ambitiously entitled the colloquium "The Bargaining Model of Social Research." I set the paper aside because it identified persons who had helped or blocked our research effort.

Some time later Professor Camillo Pellizzi asked me to submit a paper to his *Rassegna di Sociologìa*. I decided to compare my field experiences in the United States and Italy at a level of abstraction that would avoid identifying individuals (Form, 1963). In writing this paper, I found that the bargaining framework, useful for organizing the Italian experience, was inadequate for the United States. A more complex social-systems framework was more useful for the analysis of both sets of data. The ideas of this paper were subsequently elaborated and published (Form, 1971). I shall summarize them because they constitute the framework for analyzing encounters fraught with distrust in the four countries.

* I am solely responsible for interpreting the behavior of all persons identified in this account.

In most survey research, a temporary social system arises among the researchers, their sponsors, and informants. Although informants may be thought of as hosts to researchers, the latter usually want respondents to play complaint roles. When they refuse, researchers resort to various manipulations. Success depends on such factors as: the influence research sponsors have over respondents; the common and/or conflicting values of respondents, researchers, and sponsors; the relative sophistication of the three parties concerning the subject being investigated; the usefulness of the research to each party; and other specific historical and situational factors. The tactics which parties use to gain their ends in research social systems depend on their resources and skills, the complexity of the systems being studied (from simple interpersonal ones to complex interorganizational ones), whether the area under investigation threatens the parties, the type of research instruments being used, and other variables.

When researchers study organizations which normally distrust each other, they face the formidable task of building norms of trust within the research system. But they often lack sufficient knowledge of the system to design a successful access strategy. While attempting to allay the suspicion of the conflicting parties, the researchers may affect the very relationships they are trying to study. Unfortunately, text books in research methods provide little guidance on what to do in these trying circumstances. The problem of gaining research cooperation from individuals and organizations in conflict will now be analyzed in terms of these simple ideas.

The American Automobile Studies

I had studied automobile workers in Lansing, Michigan, four times over the years. Although only the fourth was part of the comparative study, I shall describe the problems of research access for the first three because they shaped the field strategy of the fourth study. All the American studies shared some characteristics with the Italian, Argentinian, and Indian studies. The interview was used to gather data from workers in their homes. The plans called for drawing samples of work-teams in various factory departments. At least the passive cooperation of management and labor was necessary to assure that workers would cooperate freely with the interviewers. Finally, workers had to be convinced that the researchers were free of union and/or management influence.

The first study of labor mobility aimed at analyzing how social origins (foreign, southern white, Negro, rural, and urban) affect the occupational allocation process (Form and Geschwender, 1962; Nosow, 1956). Oldsmobile Division of General Motors (GM) in Lansing, Michigan, was selected because it was large enough to have sufficient workers in each origin category. We approached management for permission to conduct the study and for access to their employment records. It refused. We then asked prestigious professors at the university who allegedly had good contacts with management to intervene on our behalf and ask management to draw a sample for us. Management again refused. We were then forced to change the research design and substitute an area-probability sample of workers living in the Lansing area.

The second study examined the economic, political, and social integration of workers into the union and how union participation related to neighborhood and community involvement (Form and Dansereau, 1957). A sample of workers in a large union was needed. We asked the officers of the OLDS union local for permission to do the study and for access to the seniority list in order to draw the sample. We did not approach the company, because it had earlier refused us access to its records. After obtaining initial approval from union officers, the members in an open meeting voted against endorsing the study because the faction out of power suspected that the researchers were company spies. The officers told us that only UAW headquarters in Detroit could grant permission for a study. I asked Harold Sheppard at Wayne State University, who knew officers at the International, to intervene on our behalf. He did, and we were granted permission but advised not to approach the local that had refused to cooperate.

UAW officials suggested that we approach Union Local B, an amalgamated local with seven sub-units. The officers of Local B, who had made contact with the OLDS local, became suspicious about the research. During the negotiations I realized that I had to convince them that I was sympathetic to unionism before I could gain access to the seniority list. After doing so, I was told that members of the union must be informed that the study was being sponsored jointly by the union and the university. In sum, certain compromises in research design were forced upon me: a different type of union local had to be selected; I had to demonstrate my commitment to unionism; questions in the interview had to be altered; and the study had to be defined as union-sponsored. Although these compromises were less important than those in the first study, they affected my future research in the city.

Research access problems were quite similar in the third and fourth studies. The third investigated the transfer of a department from a local Fisher Body factory to two other communities (Bloch, 1965). One phase of the third study dealt with the reallocation of workers in GM plants in three other communities, and the other phase dealt with the methods management and the union used to settle their differences. The cooperation of both union and management was needed, but the managers of the local plants refused. I sought permission to do the study from GM officials in Detroit but, after some hesitation, they also refused. The local unions agreed to cooperate and so did UAW headquarters in Detroit. A list of workers affected by the removal of the department was obtained from the union and they were interviewed. Although this study was more independent of union control than the previous one had been, it was incomplete because I could not interview foremen or gain access to company records. In short, in the three Lansing studies I was forced by the distrust of management and the union to change research plans. In the first, management denied me access to company records; in the second, the union placed me on probation; in the third, I was defined as pro-union and anti-management.

The fourth Lansing study was a replication of the Italian research. Its objective was to study the impact of technology on worker behavior on the job, and in the union, neighborhood, community, and nation. The research called for identifying workers in specific departments who had specific skills which matched those of the Italian workers. Departmental rosters had to be obtained from management because union seniority lists were not organized by departments or skill level.

I asked OLDS management for permission to do the study and for access to departmental records. Predictably, it refused, and again I sought approval from GM officials in Detroit. They refused, but directed management to give me an extended tour of the factory and covertly agreed not to discourage workers from cooperating with the study. On the tour we selected the necessary departments. The local union provided the seniority list. By a complex and devious method we were able to select an appropriate sample.

When I reviewed my field experiences in the four American studies, the Italian experience was still fresh in my mind. Research access in Italy had resembled a collective-bargaining process. I concluded that over the years in Lansing I had learned to bargain with increasing effectiveness; labor had come to accept me, and management had begun to trust me. I am now not certain that this conclusion is correct. Perhaps calm labor-management relations are more important than personal trust, for research access. In 1948, at the time of the first study, relations between the union and General Motors were tense. A long, bitter strike had recently terminated and the parties felt hostile toward one another. By the fourth study in 1963, labor-management relations were calm, and both camps had little to fear from research which required only passive cooperation.

Research Bargaining in Italy

Gaining research access in Italy was more difficult and complex than in Lansing. Italy was selected as a research site in order to study how ideological unionism affects the system involvements of workers. The problem was complicated by the presence of four unions, with different ideologies: communist, social democratic, Catholic, and "neutral." I had realized before going to Italy that the successful execution of the research would require the cooperation of the University of Turin, the company, and the four unions. Access to the university was provided by Joseph LaPalombara, who had many contacts in Italy; he wrote to an industrial sociologist at the university who agreed to sponsor me. Establishing contact with FIAT was more indirect. A colleague knew an Italian physician in the United States who had worked at FIAT and knew its general manager. I wrote the physician, described my project, and asked him to write the general manager of FIAT on my behalf. He did so. The manager wrote me of his interest in the research and asked for a brief description of the study. From the sociologist at the University of Turin I obtained the names of the chief officers of the local unions at FIAT. I wrote all of them, soliciting their cooperation in the study. The first to reply was the fascist union, which did not have a representative on the *Commissione Interna*; the second, the Catholic (CISL) union; the third, the social democratic union (UIL); and the last, the independent union (SIDA). All but the communist (CGIL) union agreed to cooperate; its officers did not reply to my first or second letter. Upon arriving in Turin, I planned first to make contact with the university, then to spend two months studying local institutions, then to involve the unions from left to right in the study and finally to present the finished research proposal to FIAT.

Initial Contacts with Unions: Ideological Dialogues

Reception at the university was cordial. Paolo Ammassari and I met top union officials in the city to explain the purpose of the research. All of them, including the communist, agreed to cooperate; they introduced us to union leaders at FIAT. In the first interview with local officials, we learned something of the history of local industrial relations, the status of current issues, and the main bargaining objectives. We explored in depth the political and ideological concerns of all unions and of management. Officials probed for our knowledge about industrial conflicts and also inquired about our own political beliefs. All officials had such detailed knowledge of the problems facing the other unions that we suspected that they had inside informants. Since no union had a list of FIAT workers, the sample had to be drawn from company files. Although all union officials had firm ideological commitments, they analyzed the local industrial and political scene rather dispassionately.

Initial Management Contacts: Indoctrination and Inquiries

The first meeting with FIAT was with officers of the Industrial Relations Department. In a formal but cordial situation, we described our broad research objectives. In the second meeting, we listed specific research needs, e.g., information on union elections, visits to selected departments, and data on the labor force. To our surprise, we were asked to specify our research aims in more detail. We were also informed that the company's cooperation in the study was contingent upon its evaluation of the material we would submit. In the meantime, officials suggested that we get to know each other better.

About a week of systematic indoctrination ensued. The company apprised us fully of its services to its employees: health, recreational, technical training, social security, housing, home for the retired, cultural events, psychological testing, and so on. We decided to submit to the company the research proposal which I had earlier submitted to the National Science Foundation and the Social Science Research Council. In the third formal meeting with industrial relations officers, the research proposal was discussed in great detail. The meeting was concluded with a meal at an exclusive restaurant, at which time we were asked to submit in writing and in detail a description of all the information we wanted.

Definition of Bargaining Goals

The delaying tactics of the company, the lack of trust, and the threat of research surveillance were irritating. We prepared a document which specified: departments to be studied, size of the sample, kinds of questions we intended to ask, people who would do the interviewing (advanced university students), needed union election data; it also requested permission to visit specific departments. During the fourth formal meeting with the industrial relations staff, we discussed each item in painstaking detail, but at the close, officials asked for a copy of the interview, the names of the interviewers, and the name of the firm which would reproduce the interview schedule. Reluctant to comply, we turned to professors at the university and to union officers for advice. They told us that we would never get permission for the study, that no one had ever received it, and that we would encounter endless obstacles.

We feared that we had no alternative but to yield to the company's demands. Perhaps we were being "captured" by FIAT: we were being indoctrinated, we were visiting company officials frequently, and we were under constant surveillance. Fearful that our relations with the unions might be endangered, we inaugurated a "balance-of-visits" scheme. Each time we visited FIAT, we visited all four unions. We also decided to assume a less compliant posture toward the company by trying to anticipate its moves and preparing to counter them. A draft of the interview was sent to the company and a request for still another meeting.

Bargaining Sessions

We waited anxiously for a week. Finally the company informed us that it had provisionally approved the study. At the fifth meeting, the company generously offered to: select the departments which fitted our specifications; provide us names of 200 people to be interviewed; help clarify some of the interview questions; perhaps eliminate some questions which would endanger the interviewers' rapport with workers; adapt some questions to the local scene; and introduce us to experienced interviewers at the School of Social Work who had worked for the Industrial Association. At this point, the industrial relations department turned us over to the personnel department to work out final arrangements.

Before our next meeting, we decided to give each union all the information we gave management. At a subsequent meeting with officers of the Personnel Department, we informed them of this action. We also assured them that we had already selected competent interviewers at the university, but they argued that the social workers were better qualified.[1] We did not agree. Management clearly wanted to eliminate questions which dealt with evaluation of the company, foremen, and the industrial relations situation. Most important, it did not want us to ask workers about their union preferences. Fortunately, one of the personnel officers was an academic man who shared our values. Although management had put him in the stressful position of monitoring our research, outside the bargaining room he gave us helpful suggestions on how to proceed.

With some important exceptions, we complied with many of management's suggestions. We demurred at employing "their" interviewers. We resisted dropping interview questions on union preference, but management insisted that the questions would damage rapport with workers and thus invalidate the study. I

[1] Later we learned that philosophy students at the university had initiated critical studies of FIAT. Obviously, the company did not want them to be interviewers.

proposed to settle the issue by a pretest, but management insisted that workers would learn about it through the grapevine. The issue remained unresolved. We could not permit the company to select the sample, force us to abandon questions on union preference, and provide interviewers who could act as company informers. Negotiations were stalled. I had about decided to abandon the research.

Mediation Stage

In desperation I asked a friend, Sr. Verdi, (pseudonym) whom FIAT management respected, to mediate. I asked him to determine whether interviewers from the School of Social Work could be trusted and whether negotiations had been going on between the school, the Industrial Association, and FIAT. Verdi found that, although FIAT had approached the association and the school, the social workers were professionally oriented and could be trusted. We permitted FIAT to introduce us to officers of the Industrial Association who, in turn, introduced us to the head of the School of Social Work (whom we had previously met). The association offered to reproduce the interviews, a generous gesture which we could hardly refuse, but which amounted to monitoring whether we would carry out our agreements with FIAT.

We urged Sr. Verdi to persuade FIAT to be more flexible on the issue of anonymity of workers to be sampled and the issue of including questions on union preference in the interview. Verdi had no contacts with the industrial relations department but, knowing other officials on the same level of authority, he pointed out that past studies of FIAT had been critical because they were done by radicals who had axes to grind. Verdi was able to convince them that a neutral and unbiased study should be permitted. Verdi also suggested that we might be willing to drop the question on union identification. After we had waited for two agonizing weeks, FIAT informed us that most of our demands could be met. The company agreed to let us draw a large sample from which we could select an anonymous subsample if we dropped the interview question on union identification.

Bargaining with the Unions

During this mediation stage, we attempted to clarify our relations with the unions. As in the past, we approached them from the Left to the Right. We asked each to provide two or three typical workers whom we could interview for the pretest. All the unions sent workers who had been members of the Commissione Interna. As trusted lieutenants, they answered every question in line with the official union position. We asked them to give their own position rather than that of the union, but they insisted that there could be no difference. After an hour of parrying, they agreed to answer the questions as they thought their fellow workers might. The encounters were not useless, because the workers criticized the ideological slant of some questions and offered other suggestions on how to improve the interview.

During these pretests we became aware of an undercurrent of hostility from representatives of the two nonradical unions: the social democratic UIL and the "nonpolitical" SIDA. We wanted both the company and the unions to approve a letter to be sent to the workers describing the study and mentioning that both had agreed to the research. During meetings with union officials, we asked them how they thought workers would respond to the question, "Which union do you identify with most?"[2] The officials' responses "scaled" perfectly, from left to right: CGIL officials insisted that the question should be asked and thought that all workers would answer it; Catholic CISL officials thought that the question should be asked and that most workers would answer it; social democratic UIL officials thought that the question was sensitive, and that some workers might not answer it; and the president of SIDA, the independent union, thought that the question was threatening and that most workers would not answer it.

Our rapport with the various unions was exactly opposite to our expectations: we had anticipated the most cooperation from the conservative unions, which now showed greatest resistance to the study. We decided to reexamine our relations with the unions—to learn about their attitudes toward us and their willingness to cooperate in the study. We suspected that the two conservative unions might be cooperating with the company in opposing some questions or even the study itself. SIDA and UIL together dominated the Commissione Interna, and they were under constant attack by the two radical unions as being pro-management.

[2] The question was needed because only 15 percent of the workers were union members. Union membership was not a requisite for voting; in fact, over 90 percent of the workers voted.

MEDIATION WITH UNIONS

Over the years the two conservative unions had allegedly been supported by the United States in their fight against CGIL. The United States Information Service (USIS) had a Labor Section in Turin which was in constant contact with the "non-communist" unions. I was on good terms with the director of USIS, with an Italian employee in the USIS Labor Section, and an American economic attaché who knew the local labor scene. I asked them to probe the attitudes of the unions toward the research and to urge them to cooperate in the study.

The reports from USIS confirmed our suspicions. SIDA, which was obligated to management, was hostile toward the study, and UIL was uncertain whether it should cooperate. USIS officials urged both to cooperate. When we next met UIL's officers, we were greeted warmly and assured of cooperation; the meeting adjourned to the bar and to rounds of toasts. SIDA officials were formal and polite; they agreed to cooperate with the study because certain issues had been "clarified." Catholic CISL had lost confidence in FIAT management and had always been inclined to cooperate. When its officers learned that a common acquaintance was related to a high CISL official in Rome, there were no more questions. Students, members of the Communist Party at the university, reassured us of CGIL's cooperation. Finally, the four unions and management had agreed to the study, the interview, and the letter to be sent out to prospective informants. Such cooperation was unheard-of in local annals.

SELECTING THE SAMPLE

To select the sample, we needed intimate knowledge of the technology of various departments. The head of automobile production, second only to the general manager, personally conducted us on a tour of the departments we wanted to inspect, including the highly secret experimental department where new models were being built. After selecting the departments, the personnel officer offered to furnish the names of the 200 workers to be interviewed. We asked for 400 names to be selected at random, so that a subsample of anonymous respondents could be selected. To our surprise he agreed, but insisted on being involved in the work. The researchers, the head of production, his assistant, and the plant psychologist sat around a table and, with a table of random numbers, selected the 400 names from departmental rosters.

Testing and Breaking Agreements

Interview-training sessions were arranged with the faculty and students at the School of Social Work. By this time we had pretested the interview on typical workers. All the interviewers were women who had had considerable experience interviewing working-class people, and we were pleased with the results of the first training session. The critical item on union preference had been dropped from the interview as a concession to management. We told the interviewers that a series of filter questions would in most cases stimulate the respondent to reveal his union preference. We felt that success in obtaining this information would test the interviewers' skill and loyalty to the researchers. We gave them a number of preliminary interviews to conduct in various sections of the city.

After three days we examined the interviews. No interviewer had obtained information on union preference, and other sensitive questions were sometimes unanswered. The interviewers admitted that they were reluctant to press for answers to sensitive questions and reported that some workers resisted being interviewed because their unions or foremen had not informed them about the study. A number of quick decisions had to be made. We decided to instruct the interviewers to ask the sensitive question on union preference. We stressed that the clinical style of interviewing was inappropriate for gathering the type of information we wanted. Uncertain whether to trust the social workers, we decided to employ "control" interviewers, in order to have a basis for comparing the quality of interviewing.[3] We then asked the unions to send some leaders in the departments under study to a meeting where we would inform them about the research, the agreement with management and the unions, the guarantees of anonymity, and the need to encourage workers to cooperate with the interviewers. The meetings were well attended. Only the independent union, SIDA, did not send a representative, but it agreed to urge workers to cooperate in the study. The interviewing was resumed and we waited anxiously for the results. The gamble paid off. Both sets of interviewers performed well. They

[3] Earlier, we wanted to hire a group of women who did interviewing commercially. The social workers, we learned, competed for the same business. Since we could not meld the two groups into a single interviewing team, members of each team were assigned different sections of the metropolitan region.

were well received by the workers, a large majority of whom answered the question on union identification.

We have used a bargaining metaphor to analyze field-work problems encountered in Turin. Distrust was so widespread among the parties that the researchers had to bargain for many things normally under their control. The areas of bargaining included: (1) union and management sponsorship; (2) contents of the letter of introduction to informants; (3) questions in the interview; (4) departments to be studied; (5) number of workers to be interviewed; (6) anonymity of the informants; (7) interviewers to be hired; (8) cooperation of the interviewers; (9) reception workers accorded the interviewers (this depended upon management's and the unions' urging workers to cooperate).

The Argentine IKA Study

The Argentine research was conducted at the Córdoba plant of Industrias Kaiser Argentina. We had hoped that the Institute of Sociology at the University of Córdoba would provide local sponsorship. Delbert C. Miller, who had previously done research in Córdoba, graciously wrote the institute on our behalf, and it agreed to help us. The plan was to send Richard E. Gale to Córdoba to do preliminary work; I would arrive just before the interviewing was to begin. Gale's task was to obtain the cooperation of management and the union. Letters describing the study were sent to the institute, IKA officials, and labor leaders before Gale departed.

When he appeared in Córdoba, he found that institute members were split in their preference for a philosophical or empirical style of research. He asked a professor who appeared to be playing a mediating role to introduce him to IKA management. After this was done, Gale was left to his own devices. He hired a research assistant on the psychology faculty, an act which irritated some institute members. Others, who regarded Gale more as a student than as a colleague, would occasionally ask him when Form would appear.

Gale early noted that he needed the cooperation of two officials at IKA: the industrial relations manager, who had access to local labor officials, and the personnel manager, who had access to employee records. Unfortunately, the two were competing for the plant manager's favor; neither was convinced that Gale was a mature scholar; both were apprehensive about cooperating with a foreign social scientist; and both were unconvinced that a study would be useful to them. Since I failed to appear and the institute's role was peripheral, Gale was held in suspicion.

Gale visited both the industrial relations manager and the personnel manager with equal regularity, so neither could accuse him of favoritism. To demonstrate the utility of social science to the personnel manager, Gale offered to study personnel turnover in the factory. Since he had nothing to offer the industrial relations manager, ties with him weakened as they improved with the personnel manager. Whenever Gale asked the industrial relations manager to be introduced to the union's officers, he was told that the time was not appropriate. Gale feared that he was being captured by management. I decided to go to Córdoba, but just before I arrived Gale had forced the issue by visiting union headquarters on his own. This provoked a meeting between Gale, the industrial relations manager, and union leaders. The latter, who wanted to demonstrate their independence from management, readily agreed to cooperate in the study. However, the industrial relations manager became hostile and may have influenced the union president to withhold cooperation.

When I arrived in Córdoba, the professor who had introduced Gale to IKA management visited me. Since conflicts within the institute were erupting and I did not want to be identified with any faction, I hurried to pay my respects to all members of the institute. They introduced me to IKA managers who gave me a tour of the plant, but at the same time they withheld permission to launch the study. Gale and I immediately completed the labor-turnover study and presented it to the personnel director, who forwarded it to the plant director. The report was well received and the personnel director agreed to cooperate in the study. We had already obtained access to the files for the turnover study and had drawn the sample. The only obstacle remaining was union agreement on a letter of introduction to be sent to prospective informants.

The president of the union, who had been working closely with management, had succeeded over the years in obtaining a decent collective bargaining agreement. However, a faction was being organized against his re-election. We had interviewed the president three times and received the impression that he had approved of the study. During the next two weeks we prepared the final interview. When we next visited the president to get his approval of the letter of introduction to be sent the workers, he was irked because we had not communicated with him for two weeks. His evasiveness in approving the letter constituted a denial of cooperation.

We had no alternative but to cultivate second-line union officials, some of whom were leading the opposition faction. We also cultivated the industrial relations manager and asked him to urge the union president to cooperate. We also asked the personnel manager to urge the industrial relations manager to go along. Apparently we convinced lower-echelon union officials that cooperation in the study would help them. With the grudging consent of the industrial relations manager, all parties finally agreed on a letter to be sent to the workers. The letter simply stated that both labor and management were aware of the study, that the research was being conducted by university professors, and that the workers could be assured of anonymity.

One consequence of all the uncertainty was that we lost our nerve about asking sensitive questions about politics. We also alienated some members of the institute either because we did not involve them sufficiently in the research or because we did not heed local deference norms. As in the other studies, in the attempt to get research clearance, we learned something about labor-management relations and union factionalism. We concluded that in this highly personalistic environment, verbal agreements remain binding only when personal relationships are constantly reinforced.

The Indian Premier Automobile Study

The last site was Premier Automobiles Limited in Bombay, India. I did not visit the plant, so the following account is based on materials provided by Baldev R. Sharma, who was completing his doctoral work under my direction. Prior to his departure Sharma and I discussed the problems of research access, and he read my diaries and reports on the subject. He promised to keep a diary of his experiences.

Before leaving he obtained the usual documents given field workers, including letters of introduction from professors in the United States and the Shri Ram Institute of India. The institute, which conducts research in industrial relations, is highly respected in India. Since it had provided some support for the project, Sharma could refer to it as a sponsor. After inspecting several plants, he decided to study the Premier plant in Bombay. His first concern was to find someone to introduce him to a plant official. He approached sociologists at a local university and presented himself as a doctoral student in the United States and a research associate in the Shri Ram Institute. Since these sociologists had not typically engaged in empirical research and had no industrial contacts, they were reluctant to sponsor him.

Sharma then approached the Tata Institute of Social Science, an agency which trained welfare workers, some of whom they sent to Premier for field-work experience. Institute members received Sharma warmly, permitted him to use its facilities, but felt they could not sponsor him at Premier. The Bombay Labor Institute, a state-supported institution, was then approached. The director tried to establish contacts at Premier, but his access was at too low a level to be useful. Sharma then approached labor union headquarters in Bombay. Although he was given considerable information about industrial relations at Premier, no official was willing to introduce him to Premier management or to local union leaders. In desperation, he approached middle management directly. He obtained an appointment with the deputy staff manager of the factory, who rebuffed him and told him that he did not have sufficient sponsorship or government authority to attempt so large a study. However, he was given an appointment to explore the subject at a later time.

Sharma then encountered an employee of the United States Information Service whom he had met as a student in the United States. He asked his American friend whether he knew anyone at Premier. The American referred him to a well-known Indian woman associated with an American educational agency. During a visit she asked Sharma to write her a letter describing precisely what he wanted to do in the plant. The letter and her recommendation were submitted to a member of the board of directors of Premier.

While awaiting a reply, Sharma returned to the deputy staff manager of the plant. This time he was well received and given permission to work with personnel records. The manager defined Sharma's presence in the plant in the same way that he defined that of social-work students from Tata Institute. He informed

Sharma that he would recommend to higher authorities that Sharma be permitted to do the research. Somewhat later the secretary to the deputy staff manager showed Sharma a letter from the company director to the deputy staff manager which simply said: "I see no harm" (in Sharma's doing the study).

The final hurdle was apparently surmounted. Middle managers presented Sharma to department supervisors and urged them to cooperate in the study. Industrial relations with the local union had become so peaceful that Sharma did not hesitate to approach local officials. After obtaining data on the history of local industrial relations, Sharma asked for union approval to interview 250 workers. An interview was arranged with higher union officials in Bombay. After they were convinced that the study had no connection with the government, the company, or the rival union, they asked local officials to give Sharma whatever help he needed for the survey.

The last problem was to overcome possible worker resistance to the interviewing. After considerable indecision, Sharma decided to interview them in the "time office" assigned to him on the factory floor. He enlisted cooperation by simply telling workers that he needed their help to write a book to complete his education. The fact that they were critical of the factory, the union, and the government suggests that they were not intimidated by being interviewed in the plant.

Interpretive Scheme

Systematic research on the politics of research access must take into account such variables as the major characteristics of the host systems, the social attributes of the research unit, how the proposed research threatens or helps those involved in the research, the resources available to participants in the research system, and how participants work out exchanges (cf. Glazer, 1972). To simplify analysis, I shall focus only on situations where the groups being studied are in conflict, are distrustful of the researcher, and have the power to withhold their cooperation. The following analysis examines the reasons why researchers are distrusted and the tactics they can use to overcome distrust.

Many conditions affect research access, but perhaps the most important is the social structure of the system being studied. Access to a community is easier than to a bureaucracy; no one has the authority to prevent researchers from studying a community, but permission must be obtained to study a bureaucracy, especially when observable research techniques are used. The more linkages external agencies have into bureaucracies the easier it is to get research access and sponsorship, simply because all linking agencies become potential sponsors. Even when permission for research is granted, some subjects are much more difficult to study than others. For example, a study of organizational control will be resisted more than will a study of the morale of randomly selected individuals in the lowest level of an organization (Form, 1971b:25-6).

It may be easier to gain research access to conflictful organizations than peaceful ones. In conflict situations, second-level officials may cooperate with researchers more readily than will top officials, because people at the lower levels keenly resent the denial of power. Sometimes the weakest parties cooperate most with researchers because they have the least to lose. Detailed knowledge of the conflict then becomes the researcher's most useful tool for persuading reluctant respondents to release information. Where it is difficult to contain knowledge of intra-system conflicts, parties eagerly release information to researchers to avoid giving the opposition greater credibility. However, conflict is structured differently in different societies, and researchers must take this into account in designing access strategies. For example, in the United States industrial conflict tends to be situational and focussed on issues known only to the parties, while in Europe industrial conflict is often part of general political conflict. The privacy inherent in American industrial relations makes research access more difficult than it is in Europe.

Another important variable that affects research access to conflict situations is the extent to which parties feel that the research findings will change their power relations. Resistance is high when informants feel that the revelation of knowledge will weaken their position in the system. Researchers, in turn, are most powerless (i.e., most dependent on respondents) when they are ignorant of the system under study. Informants are then in a position to manipulate researchers for their private ends. In this situation researchers must rely on the strength of their sponsors, their own prestige or reputation, and their ability to use whatever knowledge they have accumulated. Whenever ideological conflict in the system under study is high, how the parties label the researchers ideologically will affect their cooperation. With this brief background, we shall review the cases under study.

Bureaucratic Distrust: OLDS

In studies of automobile workers, I was seeking cooperation from companies and unions which had centralized authority structures. Access tactics useful in community research were not effective. In bureaucracies, managers have enormous power; if they distrust research, it will not be done. Researchers meet most resistance when they study the top management where decisions are being made. Managers have power to deny researchers access to the organization, and they often do so because they realize that the research may expose how they use their power and authority.

In Lansing, I was never able to study areas over which management had complete control, but I was permitted to study areas where management shared power with labor. Management denied me permission to study the behavior of work groups, the occupational allocation of workers according to their social origins, and the transfer of the cut-and-sew department to other cities. It resisted my requests with impunity; it had nothing to gain by permitting an outsider to examine its decisions. Management did permit others to study the morale of individual workers (selected at random), but it did not permit anyone to study its personnel decisions. For over twenty years, I could not persuade the OLDS management to trust me. Yet it did not use all its resources to stop my research; e.g., it never advised workers to refuse to be interviewed nor did it ask the unions to honor the contract clause which forbids outsiders access to the seniority list. Possibly management tolerated this violation because I did not study "important" issues and because I did not feed the union damaging information about the company. However, research can threaten management; e.g., during a strike, information on derogatory attitudes of employees toward the company could be released.

Why did local unions cooperate? First, I demonstrated my loyalty to unionism by letting them know that I had contacts with top union leaders in Detroit. Second, I did not propose to examine internal union problems; I did not threaten the authority of officials, who fear studies of union democracy as much as managers fear studies of personnel decisions. Third, and perhaps most important, the unions conformed to the tendency for the weaker parties in a conflict to cooperate with outsiders who might help them. (Form, 1971b:31).

Ubiquitous Conflict, Ideology, and Bargaining: FIAT

The FIAT case draws attention to the relevance of social conflict and political ideologies for research access. Clearly, like OLDS, FIAT could have refused to sanction the research, ending it immediately; FIAT had the only list of workers needed to draw a sample. I can only speculate why the company decided to cooperate: as a gesture of good will, a public relations gesture, or the hope of a favorable study. Once the company decided to cooperate, two environmental factors explain much of its behavior: (a) general distrust of organizations over which it had no control; (b) conflict among the unions and between them and management. For a number of years FIAT had fought the communist-dominated CGIL. After the company had reduced the union's influence in the plant, FIAT was able to dominate labor-management relations (Ornati, 1963). Management profited from the continuing union rivalries which tipped the balance of power in its favor.

Once FIAT agreed to the research, two things became inevitable. First, the company had to exert maximum surveillance over the research in order to prevent a critical study, and, second, researchers were given bargaining power because they too could operate to advantage in the fractionalized environment. In the ensuing bargaining, the company had the clear advantage; it had more resources, it could withdraw from the agreement; it had contacts with the School of Social Work and the Industrial Association, and it dominated the weak unions. Since withdrawing from the research might advertise that the company had something to hide, it chose to monitor rather than oppose the research.

The behavior of the unions is easier to explain. As weak parties in industrial bargaining, the radical unions had nothing to lose and possibly something to gain from the research. They were convinced that a neutral scientist would produce findings which would support their view that management's policies were repressive. The hesitation of conservative unions in power to cooperate in the research reflected their ties to management; aggravating the company by pleasing the researchers was not a risk worth taking. Yet to withhold cooperation would give credibility to charges that conservative unions were pro-management. Their decision to cooperate was made easier by the company's ultimate consent to go along and by the good offices of United States agencies.

The main bargaining resource available to the researchers was their ability to take advantage of conflict. As long as a single party was willing to cooperate, reluctant parties found it difficult to provide convincing reasons why they should not. To refuse would be a sign of weakness or even collusion with management. The status of a foreign educator and sponsorship from the local university and United States agencies were other resources favorable to the researcher.

Distrust and Opportunism: IKA

Inability to gain research access in Argentina was due as much to conflict within management and labor as to conflict between the two. Distrust and opportunism seemed to be a structural feature of the culture. Only in Argentina did factions within management and within the union find the threat of cooperating with researchers useful to their own politics. Perhaps this condition was responsible for the fragile nature of the agreements with the researchers, who found it necessary to reinforce their personal contacts constantly. Once an agreement was made, it had to be executed immediately because it might not endure. Sponsorship from the outside was not sufficiently strong to overcome pervasive distrust within and between the parties.

External Linkages and Personal Sponsorship: PAL

The Indian research environment was distinctive. Labor and management had relatively few linkages to the external community, and their internal relationships were informal but hierarchical. This situation made it difficult for Sharma to find suitable research sponsors. Distrust of outsiders was so strong that research access could only be achieved by finding persons who had strong particularistic ties with company officials. Once permission for the study was given by a powerful company official, the researcher could count on the cooperation of the lower echelons.

The same situation applied for the union. Overcoming distrust was a problem only during the early stages of field research in India, while in Italy and Argentina distrust seemed endemic and ubiquitous.

Persistent Issues

These tentative observations on the relevance of social structure to research access leave two important questions: Were the problems encountered due to societal differences or to situational factors? Can a strategy of research access be developed on the basis of knowledge about the social systems being studied? My limited experience suggests that the problems reported here were not unique or situational. In American industrial relations, labor and management operate rather independently except for formal bargaining; they are not concerned about each other's internal operations (Dunlop, 1970:94-128). In Italy, industrial relations are part of a larger political scene, so the parties are concerned about issues in various contexts (LaPalombara, 1957: 71-104). The Argentinian government was so repressive to labor that industrial relations perpetually involved the demand for legitimacy on the part of labor (Alexander, 1962:210-22). Unions in India are weak bargaining agencies, more interested in national politics than in what is going on in management and in the factory (Myers, 1959:19-74).

Research can be sponsored by unions in the United States without arousing management's hostility, and it can be sponsored by management in India without arousing undue suspicion from labor. In the factionalized industrial-relations climate of Italy, researchers encounter great difficulty in maintaining a neutral stance: their activities are assigned ideological significance by all parties. The unsatisfied demand for recognition on the part of labor in Argentina results in the withholding of cooperation until its claim for status is validated. Lack of confidence in organizational decisions adds to the complications (Fillol, 1961). Tactics of research access can be improved by taking account of such social structural differences in industrial relations.

But researchers face a dilemma when studying social systems about which they have little previous knowledge. It is precisely the absence of this knowledge that motivates them to undertake the research. The challenge they face in the field is to obtain reliable knowledge about social systems rapidly enough to anticipate access problems and to design tactics to meet them. The inability of many researchers to do this accounts for the high mortality of crossnational studies and the shifting of research problems in midstream.

Unfortunately, knowledge learned from failures to gain access usually remains private. It is now possible to design comparative studies to accumulate systematic knowledge on research as a social process. Alternative field strategies can be planned and records kept on the outcomes. Comparative research offers sociologists quasi-experimental conditions to study the effects of local social systems on field strategy; the problems, instrumentation, and other research variables are similar, but the host communities differ. Materials commonly found in field-methods courses have not been tested in the context of different socio-cultural settings. Studies must be designed to show how different organizational settings of the research (bureaucracies, institutions, communities) in different societies affect research as a social process. In the meantime, social scientists should publish their field experiences to enable other scholars to profit from their errors.

References

Abegglen, James.
1959 The Japanese Factory. Cambridge, Massachusetts: M.I.T.

Alexander, Robert J.
1962 Labor Relations in Argentina, Brazil and Chile. New York: McGraw-Hill.

Allardt, Erik.
1965 "Samhällsstruktur och sociala spänningar. Helsingfors: Söderstroms." Cited in Joachim Israel, Alienation: From Marx to Modern Sociology. Boston: Allyn and Bacon, 1971.

Almond, Gabriel A., and Sidney Verba.
1965 The Civic Culture. Boston: Little, Brown.

Ammassari, Paolo.
1964 "Worker satisfaction and occupational life, a study of the automobile worker in Italy." Ph.D. dissertation. Michigan State University, East Lansing, Michigan.

———.
1969 "The Italian blue-collar worker." International Journal of Comparative Sociology 10 (March and June): 3-21.

Angell, Robert Cooley.
1951 The Moral Integration of American Cities. Chicago: University of Chicago Press.

Aronowitz, Stanley.
1973 False Promises: The Shaping of American Working Class Consciousness. New York: McGraw-Hill.

Armer, Michael, and Allan Schnaiberg.
1973 "Measuring individual modernity: A near myth." Pp. 249-281 in Michael Armer and Allen D. Grimshaw (eds.), Comparative Social Research: Methodological Problems and Strategies. New York: Wiley.

Balandier, Georges, and Paul Mercier.
1962 "Le travail dans regions en voie d'industrialisation." Pp. 282-304 in Georges Friedmann and Pierre Navalle (eds.), Traité de Sociologie du Travail. Vol. 2. Paris: Armand Colin.

Baranson, Jack.
1969 Automotive Industries in Developing Countries. Baltimore, Maryland: Johns Hopkins Press.

Barbichon, Guy.
1967 "La mobilité comme objet d'action. Une illustration: Le transfèr des agriculteurs dans l'industrie, en Italie." Sociologie du Travail 4 (October-December):421-437.

Bauman, Zygmunt.
1972 Between Class and Elite. Manchester: Manchester University Press.

Beijer, G.
1963 Rural Migrants in Urban Setting. The Hague: Martinus Nijhoff.

Bell, Daniel.
1973 The Coming of Post-Industrial Society: A Venture in Forecasting. New York: Basic Books.

Bell, Wendell, and Marion D. Boat.
1957 "Urban neighborhoods and informal social relations." American Journal of Sociology 67 (January):391-400.

Belshaw, Cyril S.
1965 Traditional Exchange and Modern Markets. Englewood Cliffs, New Jersey: Prentice-Hall.

Bendix, Reinhard.
1956 Work and Authority in Industry. New York: Wiley.

Bendix, Reinhard, and Seymour Martin Lipset.
1952 "Social mobility and occupational career patterns." American Journal of Sociology 57 (January):366-374.

Bennett, John W., and Iwao Ishino.
1963 Paternalism in the Japanese Economy. Minneapolis: University of Minnesota.

Berger, Bennett M.
1960 Working-Class Suburb. Berkeley: University of California Press.

Berger, Peter (ed.).
1964 The Human Shape of Work. New York: Macmillan.

Blau, Peter M.
1964 "The research process in the study of the dynamics of bureaucracy." Pp. 24-32 in Phillip E. Hammond (ed.), Sociologists at Work. New York: Basic Books.

Blau, Peter M., and Otis Dudley Duncan.
1967 The American Occupational Structure. New York: Wiley.

Blauner, Robert.
1960 "Work satisfaction and industrial trends." Pp. 339-360 in Walter Galenson and Seymour Martin Lipset (eds.), Labor and Trade Unionism. New York: Wiley.

———.
1964 Alienation and Freedom: The Factory Worker and His Industry. Chicago: University of Chicago Press.

Bloch, Heinz.
1965 Some Sociological and Economic Consequences of a Departmental Shutdown. Winterthur, Switzerland: Keller.

Blood, Milton R., and Charles L. Hulin.
1967 "Alienation, environmental characteristics, and worker reponses." Journal of Applied Psychology 51 (June):284-290.

Blumberg, Paul.
1968 Industrial Democracy: The Sociology of Participation. London: Constable.

Blumer, Herbert.
1960 "Early industrialization and the laboring class." Sociological Quarterly 1 (January):5-14.

Boeke, J. H.
1942 The Structure of the Netherlands Indian Economy. New York: Institute of Pacific Relations.

Bonazzi, Giuseppe.
1964 Alienazione e Anomìa nella Grande Industria. Milano: Edizioni Avanti!

Booth, Alan, and John Edwards (eds.).
1972 Social Participation in Urban Society. Morristown, New Jersey: Schenkman.

Breeze, Gerald (ed.).
1969 The City in Newly Developing Countries. Englewood Cliffs, New Jersey: Prentice-Hall.

Briones, Guillermo.
1963 "Training and adaptation of the labor force in the early stages of industrialization." International Social Science Journal 15 (No. 4):571-580.

Browning, H. L.
1971 "Migrant selectivity and the growth of large cities in developing societies." Pp. 273-314 in Study Committee of the Office of the Foreign Secretary of the National Academy of Sciences with Support of the Agency for International Development (eds.), Rapid Population Growth. Baltimore: Johns Hopkins.

Bureau of Labor Statistics.
1961 Labor in India, Report No. 188. Washington, D.C.: U.S. Department of Labor. (April)

Burns, Tom, and G. M. Stalker.
1961 The Management of Innovation. Chicago: Quadrangle Books.

Caplow, Theodore.
1954 The Sociology of Work. Minneapolis: University of Minnesota Press.

Carocci, Giovanni.
1960 Inchiesta alla Fiat. Firenze: Parenti Editore.

Carr, Leslie G.
1971 "The Srole items and acquiescence." American Sociological Review 36 (April):287-293.

Chamberlain, Neil W., and Donald E. Cullen.
1971 The Labor Sector. New York: McGraw-Hill.

Chaplin, David.
1967 The Peruvian Industrial Labor Force. Princeton, New Jersey: Princeton University Press.

Chinoy, Ely.
1955 Automobile Workers and the American Dream. Garden City, New York: Doubleday.

Clack, Garfield.
1967 Strikes and Dissatisfaction with Assembly Line Work in Car Factories. Cambridge Opinion 45. Department of Applied Economics, University of Cambridge.

Clark, Colin.
1957 The Conditions of Economic Progress. London: Macmillan.

Cohen, Albert K.
1965 "The sociology of the deviant act." American Sociological Review 30 (February):5-14.

Cole, Robert E.
1971 Japanese Blue Collar: The Changing Tradition. Berkeley: University of California.

Coleman, James S.
1968 "Modernization: Political aspects." Pp. 395-402 in David L. Sills (ed.), International Encyclopedia of the Social Sciences. New York: Macmillan and The Free Press.

Conde, Roberto C.
1965 "Problemas del creciamiento industrial." Pp. 59-84 in Torcuato S. di Tella, Gino Germani, Jorge Graciarena, y

colaboradores (eds.), Argentina, Sociedad de Masas. Buenos Aires: Editorial Universitaria de Buenos Aires.

Conklin, George H.
1973 "Factory workers and social change in a smaller Indian city." Paper delivered at the Annual Meeting of the Association for Asian Studies, Palmer House, Chicago.

Consejo Nacional de Desarrollo.
1965 Presidencia de la Nación, Distribución del Ingres Cuentos Nacionales en la Argentina. Vol. v of Investigación conjuntes CONADE-CEPAL, Población y Remuneraciones. Buenos Aires.

Cottrell, Fred.
1955 Energy and Society. New York: McGraw-Hill.

Craven, Paul, and Barry Wellman.
1974 "The network city." Pp. 57-84 in Marcia Pelly Effrat (ed.), The Community. New York: The Free Press.

Curtis, James.
1971 "Voluntary association joining: A cross-national comparative note." American Sociological Review 36 (October):872-880.

Darling, Birt.
1950 City in the Forest: The Story of Lansing. New York: Stratford House.

Davis, Stanley M., and L. W. Goodman (eds.).
1972 Workers and Managers in Latin America. Lexington, Massachusetts: Heath.

Dean, Dwight.
1961 "Alienation: Its meaning and measurement." American Sociological Review 26 (October):753-758.

Dean, Lois R.
1954 "Social integration, attitudes and union activity." Industrial and Labor Relations Review 8 (October):48-58.

Dirección General de Estadistica, Censos, e Investigaciones.
1967 Cuadro 11. Córdoba: Faculdad de Ciencias Economicas, Consejo Nacional de Desarrollo.

Divisione Lavoro e Statistica.
1959 Annuario Statistico. Città di Torino.

Dore, Ronald.
1973 British Factory-Japanese Factory. Berkeley and Los Angeles: University of California Press.

Dotson, Floyd.
1951 "Patterns of voluntary association among urban working-class families." American Sociological Review 16 (October):687-693.

Dubin, Robert.
1956 "Industrial workers' worlds: A study of the 'central life interests' of industrial workers." Social Problems 3 (January): 131-142.

Dunlop, John T.
1970 Industrial Relations Systems. Carbondale, Illinois: Southern Illinois University Press.

Durkheim, Emile.
1964 The Division of Labor in Society. Trans. George Simpson. Glencoe, Illinois: The Free Press.

Edelman, Murray.
1969 "The conservative political consequences of labor conflict." Pp. 163-176 in Gerald G. Somers (ed.), Essays in Industrial Relations Theory. Ames, Iowa: Iowa State University Press.

Eisenstadt, S. N.
1964 "Social change, differentiation and evolution." American Sociological Review 29 (June):375-386.

Ellul, Jacques.
1967 The Technological Society. Trans. John Wilkinson. New York: Vintage.

Farace, R. Vincent.
1966 "A study of mass communication and national development." Journalism Quarterly 43 (Summer):305-313.

Faunce, William A.
1968 Problems of an Industrial Society. New York: McGraw-Hill.

Faunce, William A., and William H. Form (eds.).
1969 Comparative Perspectives on Industrial Society. Boston: Little-Brown.

Faunce, William A., and M. Joseph Smucker.
1966 "Industrialization and community status structure." American Sociological Review 31 (June):390-399.

Feldman, Arnold S., and Wilbert E. Moore.
1960 "The market." Pp. 41-58 in Wilbert E. Moore and Arnold S. Feldman (eds.), Labor Commitment and Social Change in Developing Areas. New York: Social Science Research Council.

Feuer, Lewis.
1963 "What is alienation: The career of a concept." Pp. 127-148 in Maurice Stein and Arthur Vidich (eds.), Sociology on Trial. Englewood Cliffs, New Jersey: Prentice-Hall.

Fillol, Tomas Roberto.
1961 Social Factors in Economic Development: The Argentine Case. Cambridge, Massachusetts: M.I.T. Press.

Finifter, Bernard.
1973 "Replication strategies in social science." East Lansing: Michigan State University. (Unpublished manuscript.)

Foley, Donald L.
1950 "The use of facilities in a metropolis." The American Journal of Sociology 56 (November):238-246.

Form, William H.
1959 "Organized labor's place in the community power structure." Industrial and Labor Relations Review 12 (July):526-539.

———.
1963 "Sulla sociologìa della ricerca sociale." Rassegna Italiana di Sociologìa 4 (Settembre):463-481.

———.
1969 "Occupational and social integration of automobile workers in four countries: A comparative study." International Journal of Comparative Sociology 10 (March-June):95-116.

———.
1971a "The accommodation of rural and urban workers to industrial discipline and urban living: A four-nation study." Rural Sociology 36 (December):488-508.

———.
1971b "The sociology of social research." Pp. 3-42 in Richard O'Toole (ed.), The Management, Organization, and Tactics of Social Research. Boston: Schenkman.

———.
1972. "Technology and social behavior of workers in four countries: A sociotechnical perspective." American Sociological Review 37 (December):727-738.

———.
1973a "Job vs. political unionism: A cross-national comparison." Industrial Relations 12 (May):224-238.

———.
1973b "Autoworkers and their machines: A study of work, factory,

and job satisfaction in four countries." Social Forces 51 (December):1-15.

———.
1973c "The internal stratification of the working class: System involvements of autoworkers in four countries." American Sociological Review 38 (December):697-711.

———.
1974a "Field problems in comparative research: The politics of distrust." Pp. 83-117 in J. Michael Armer and Allen D. Grimshaw (eds.), Methodological Problems in Comparative Research. New York: Wiley.

———.
1974b "Automobile workers in four countries: The relevance of system participation for working-class movements." British Journal of Sociology 25 (December):442-460.

———.
1975a "The social construction of anomie: A four-nation study of industrial workers." American Journal of Sociology 80 (March):1165-1191.

———.
1975b "The internal stratification of the working class: A reanalysis." American Sociological Review 40 (August):532-536.

Form, William H., and H. Kirk Dansereau.
1957 "Union member orientations and patterns of social integration." Industrial and Labor Relations Review 11 (October): 3-12.

Form, William H., and Geschwender, James A.
1962 "Social reference basis of job satisfaction: The case of manual workers." American Sociological Review 27 (April):228-237.

Form, William H., and Warren L. Sauer.
1963 "Labor and community influentials: A comparative study of participation and imagery." Social Forces 17 (October):3-19.

Fossati, Antonio.
1951 "The fiftieth anniversary of the Fiat and the future of Italian industry." Pp. 219-272 in "Fiat" A Fifty Years' Record. Verona: Arnoldo Mondadori.

Fox, Alan.
1974 Beyond Contract: Work, Power and Trust Relations. London: Farber and Farber.

Freeman, Howard E., and Morris Showel.
1951 "Differential political influence of voluntary associations." Public Opinion Quarterly 15 (Winter):703-714.

Freeman, Howard E., Edwin Novak, and Leo G. Reeder.
1957 "Correlates of membership in voluntary associations." American Sociological Review 22 (October):528-533.

Freedman, Ronald, and Deborah Freedman.
1956 "Farm-reared elements in the non-farm population." Rural Sociology 21 (March):50-61.

Friedlander, Frank.
1966 "Importance of work versus nonwork among socially and occupationally stratified groups." Journal of Applied Psychology 50 (December):437-441.

Friedmann, Georges.
1955 Industrial Society: The Emergence of the Human Problems of Automation. Edited by Harold L. Sheppard. Glencoe, Illinois: The Free Press.

Fromm, Eric.
1955 The Sane Society. New York: Holt, Rinehart, and Winston.

Froomkin, Joseph, and A. J. Jaffe.
1953 "Occupational skill and socioeconomic structure." American Journal of Sociology 59 (July):42-48.

Fullan, Michael.
1970 "Industrial technology and worker integration in the organization." American Sociological Review 36 (December): 1028-1039.

Fuller, Varden.
1970 Rural Worker Adjustment to Urban Life. Ann Arbor: Institute of Labor and Industrial Relations, The University of Michigan-Wayne State University, the National Manpower Policy Task Force.

Gale, Richard P.
1969 "Industrial development and the blue-collar worker in Argentina." International Journal of Comparative Sociology 10 (March and June):117-150.

Galbraith, John Kenneth.
1967 The New Industrial State. Boston: Houghton Mifflin.

Gamson, William A.
1966 "Reputation and resources in community politics." American Journal of Sociology 72 (September):121-31.

Germani, Gino.
1966 "Social and political consequences of mobility." Pp. 364-394 in Neil J. Smelser and Seymour Martin Lipset (eds.), Social Structure and Mobility in Economic Development. Chicago: Aldine.

Gerstl, Joel E.
1961 "Determinants of occupational community in high status occupations." Sociological Quarterly 2 (January):37-49.

Giddens, Anthony.
1973 The Class Structure of Advanced Societies. New York: Harper and Row.

Gist, Noel P.
1968 "Urbanism in India." Pp. 22-32 in Sylvia Fleis Fava (ed.), Urbanism in World Perspective. New York: Crowell.

Glazer, Myron.
1972 The Research Adventure. New York: Random House.

Glenn, Norval D., and Jon P. Alston.
1968 "Cultural distances among occupational categories." American Sociological Review 33 (June):365-382.

Goldthorpe, John H., David Lockwood, Frank Bechhofer, and Jennifer Platt.
1968 The Affluent Worker: Political Attitudes and Behavior. London: Cambridge University Press.

———.
1969 The Affluent Worker in the Class Structure. London: Cambridge University Press.

Goodman, Leo A.
1971 "The analysis of multidimensional contingency tables: Stepwise procedures and direct estimation methods for building models for multiple classifications." Technometrics 13 (February):33-61.

———.
1972 "A general model for analysis of surveys." American Journal of Sociology 77 (May):1035-1086.

———.
1973 "Causal analysis of data from panel studies and other kinds of surveys." American Journal of Sociology 78 (March): 1135-1191.

Gore, M.S.
1970 Immigrants and Neighborhoods: Two Aspects of Life in a Metropolitan City. Bombay: Sion. Tata Institute of Social Sciences.

Government of India.
1960 Annual Survey of Industries—1960. Vol. IX.

Gregory, Peter.
1960 "The labor market in Puerto Rico." Pp. 136-172 in Wilbert E. Moore and Arnold S. Feldman (eds.), Labor Commitment and Social Change in Developing Areas. New York: Social Science Research Council.

Gurin, Gerald, Joseph Veroff, and Sheila Feld.
1960 Americans View Their Mental Health. New York: Basic Books.

Gusfield, Joseph R.
1967 "Tradition and modernity: Misplaced polarities in the study of social change." American Journal of Sociology 72 (January):351-362.

Haller, Archibald O., Donald B. Holsinger, and Helcio Ulhoa Saraiva.
1972 "Variations in occupational prestige hierarchies: Brazilian data." American Journal of Sociology 77 (March):941-956.

Hamilton, Richard F.
1964 "The behavior and values of skilled workers." Pp. 43-57 in Arthur B. Shostak and William Gomberg (eds.), Blue-Collar World. Englewood Cliffs, New Jersey: Prentice-Hall.

———.
1965 "Skill level and politics." Public Opinion Quarterly 31 (Winter):390-399.

———.
1967 Affluence and the French Worker in the Fourth Republic. Princeton: Princeton University Press.

Harbison, F. H., and E. W. Burgess.
1954 "Modern management in western Europe." American Journal of Sociology 60 (July):15-23.

Harbison, Frederick, and Charles A. Myers.
1964 Education, Manpower, and Economic Growth. New York: McGraw-Hill.

Hausknecht, Murray.
1962 The Joiners: A Sociological Description of Voluntary Association Membership in the United States. New York: Bedminster.

———.
1964 "The blue-collar joiner." Pp. 207-215 in Arthur B. Shostak and William Gomberg (eds.), Blue-Collar World: Studies of the American Worker. Englewood Cliffs, New Jersey: Prentice-Hall.

Hawley, Amos H.
1971 Urban Society. New York: Ronald Press.

Higgins, Benjamin.
1956 "The dualistic theory of underdeveloped areas." Economic Development and Cultural Change 4 (January):99-115.

Hodge, Robert W., Donald J. Treiman, and Peter H. Rossi.
1966 "A comparative study of occupational prestige." Pp. 309-332 in Reinhard Bendix and Seymour Martin Lipset (eds.), Class Status and Power. New York: Free Press.

Horowitz, Irving Louis.
1972 Foundations of Political Sociology. New York: Harper and Row.

Hoselitz, Bert F.
1955 "The city, the factory, and economic growth." American Economic Review 45 (May):166-184.

———.
1962 "The development of the labor market in the process of economic growth." Transactions of the Fifth World Congress of Sociology 2: The Sociology of Development:51-71.

———.
1964 "Social stratification and economic development." International Social Science Journal 16 (No. 2):237-251.

Hoselitz, Bert F., and Wilbert E. Moore (eds.).
1963 Industrialization and Society. The Hague: UNESCO-Mouton.

Huber, Joan, and William H. Form.
1973 Income and Ideology: An Analysis of the American Political Formula. New York: The Free Press.

Hyman, Herbert H., and Charles R. Wright.
1971 "Trends in voluntary association membership of American adults: Replication based on secondary analysis of national sample surveys." American Sociological Review 36 (April): 191-206.

(de)Imaz, José Luis.
1970 Los Que Mandan. Trans. and with an Introduction by Carlos A. Astiz. Albany: State University of New York Press.

Inkeles, Alex.
1960 "Industrial man: The relation of status to experience, perception and value." American Journal of Sociology 66 (July): 1-31.

———.
1966 "The modernization of man." Pp. 138-150 in Myron Weiner (ed.), Modernization: The Dynamics of Growth. New York: Basic Books.

———.
1969a "Making men modern: On the causes and consequences of individual change in six developing countries." American Journal of Sociology 75 (September):208-225.

———.
1969b "Participant citizenship in six developing countries." American Political Science Review 63 (December):1120-1141.

Inkeles, Alex, and Peter H. Rossi.
1956 "National comparisons of occupational prestige." American Journal of Sociology 61 (January):329-339.

IRES.
1959 Panorama Economico e Sociale della Provincia di Torino. Torino.

Israel, Joachim.
1971 Alienation: From Marx to Modern Sociology. Boston: Allyn and Bacon.

ISTAT.
1964 Decimo Censimento della Popolazione. Vol. II. Roma.

Iutaka, S.
1963 "Mobilidada social e opportunidades educacionais en Buenos Aires e Moteviden: Una análise comparativa." America Latina 6 (April):21-39.

Jackman, Mary.
1972 "The political orientation of the socially mobile in Italy: A re-examination." American Sociological Review 37 (April): 213-222.

Janowitz, Morris.
1952 The Community Press in an Urban Setting. Glencoe, Illinois: The Free Press.

———.
1970 Political Conflict: Essays in Political Sociology. Chicago: Quandrangle Books.

Jaques, Elliott.
1952 The Changing Culture of a Factory. New York: Dryden.

Jencks, Christopher, Marshall Smith, Henry Acland, Mary Jo Bane, David Cohen, Herbert Gintis, Barbara Heyns, and Stephan Michelson.
1972 Inequality. New York: Basic Books.

Kahl, Joseph.
1968 The Measurement of Modernism. Austin: University of Texas.

Karsh, Bernard, and Robert E. Cole.
1968 "Industrialization and the convergence hypothesis: Some aspects of contemporary Japan." Journal of Social Issues 24 (October):45-64.

Keller, Suzanne.
1968 The Urban Neighborhood. New York: Random House.

Kerr, Clark, and Abraham Siegel.
1954 "The interindustry propensity to strike—an international comparison." Pp. 189-212 in Arthur Kornhauser, Robert Dubin, and Arthur M. Ross (eds.), Industrial Conflict. New York: McGraw-Hill.

Kerr, Clark, John T. Dunlop, Frederick H. Harbison, and Charles A. Myers.
1960 Industrialism and Industrial Man. Cambridge, Massachusetts: Harvard University Press.

Key, William H.
1968 "Rural-urban social participation." Pp. 305-312 in Sylvia Fleis Fava (ed.), Urbanism in World Perspective. New York: Crowell.

Knupfer, Genevieve.
1947 "Portrait of the underdog." Public Opinion Quarterly 11 (Spring):103-114.

Kornhauser, Arthur.
1965 Mental Health of the Industrial Worker. New York: Wiley.

Kriesberg, Louis.
1963 "Entrepreneurs in Latin America and the role of cultural and situational processes." International Social Science Journal 15 (No. 4):581-594.

Kuczynski, Jürgen.
1971 The Rise of the Working Class. Trans. C.T.A. Ray. New York: McGraw-Hill.

Labor Bureau.
1965 Statistics of Factories, 1962. Government of India.

Lambert, Richard D.
1963 Workers, Factories, and Social Change in India. Princeton: Princeton University Press.

Landsberger, Henry A., Manuel Barrera, and Abel Toro.
1964 "The Chilean labor union leader: Attitudes toward labor relations." Industrial and Labor Relations Review 17 (April): 399-420.

LaPalombara, Joseph.
1957 The Italian Labor Movement: Problems and Prospects. Ithaca, New York: Cornell University Press.

———.
1965 "Italy: Fragmentation, isolation, alienation." Pp. 282-330 in Lucian W. Pye and Sidney Verba (eds.), Political Culture and Political Development. Princeton: Princeton University Press.

Lee, Alfred McClung.
1972 "An obituary for alienation." Social Problems 20 (Summer): 121-127.

Leggett, John C.
1963a "Uprootedness and working-class consciousness." American Journal of Sociology 68 (May):682-692.

———.
1963b "Working-class consciousness, race, and political choice." American Journal of Sociology 69 (September):171-186.

———.
1964 "Economic insecurity and working-class consciousness." American Sociological Review 29 (April):226-234.

Lenin, Nikolai.
1943 Selected Works. Vol. X. New York: International Publishers.

Lerner, Daniel.
1968 "Modernization: Social aspects." Pp. 386-395 in David L. Sills (ed.), International Encyclopedia of the Social Sciences. New York: The Macmillan Co. and The Free Press.

Lipset, Seymour Martin.
1960 Political Man: The Social Bases of Politics. Garden City, New York: Doubleday.

———.
1970 Revolution and Counterrevolution. Garden City, New York: Doubleday.

Lipset, Seymour Martin, and Reinhard Bendix.
1959 Social Mobility in Industrial Society. Berkeley: University of California Press.

Lipset, Seymour Martin, Martin A. Trow, and James S. Coleman.
1956 Union Democracy. Glencoe, Illinois: The Free Press.

Lipsitz, Lewis.
1964 "Work life and political attitudes: A study of manual workers." American Political Science Review 68 (December): 951-962.

Litt, Edgar.

1963 "Civic education, community norms, and political indoctrination." American Sociological Review 28 (February):69-75.

———.

1966 "The politics of a cultural minority." Pp. 105-122 in M. Kent Jennings and L. Harmon Zeigler (eds.), The Electoral Process. Englewood Cliffs, New Jersey: Prentice-Hall.

Lopreato, Joseph.

1967a Peasants No More. San Francisco: Chandler Publishing Company.

———.

1967b "Upward social mobility and political orientation." American Sociological Review 32 (August):586-592.

Lopreato, Joseph, and Lawrence E. Hazelrigg.

1970 "Intragenerational vs. intergenerational mobility in relation to sociopolitical attitudes." Social Forces 49 (December): 200-210.

Mackenzie, Gavin.

1973 The Aristocracy of Labor: The Position of Skilled Craftsmen in the American Class Structure. London: Cambridge University Press.

(de) Man, Henri.

1929 Joy in Work. Trans. Eden and Cedar Paul. London: Allen and Unwin.

Marcuse, Herbert.

1964 One-Dimensional Man. Boston: Beacon Press.

Marsh, Robert M.

1967 Comparative Sociology. New York: Harcourt, Brace, and World.

Marx, Karl.

1956 Selected Writings in Sociology and Social Philosophy. Edited by T. B. Bottomore and Maximilien Rubel. New York: McGraw-Hill.

———.

1963 Early Writings. Trans. and Edited by T. B. Bottomore. New York: McGraw-Hill.

Marx, Karl, and Friedrich Engels.

1959 "Manifesto of the Communist Party." Pp. 1-41 in Lewis S. Feuer (ed.), Basic Writings on Politics and Philosophy. New York: Anchor Books.

Mayo, Elton.
1945 The Social Problems of an Industrial Civilization. Cambridge, Massachusetts: Harvard University Press.

McClosky, Herbert, and John H. Schaar.
1965 "Psychological dimensions of anomy." American Sociological Review 30 (February):14-40.

McDill, Edward, and Jeanne Clare Ridley.
1962 "Status, anomia, political alienation, and political participation." American Journal of Sociology 68 (September):205-213.

Mehta, R. L.
1960 A Study of the Strike in the Premier Automobiles Ltd. from the Point of View of the Code of Discipline. New Delhi: Ministry of Labour and Employment.

Meier, Dorothy L., and Wendell Bell.
1959 "Anomia and differential access to the achievement of life goals." American Sociological Review 24 (April):189-207.

Meissner, Martin.
1969 Technology and the Worker. San Francisco: Chandler Publishing Co.

———.
1971 "The long arm of the job: A study of work and leisure." Industrial Relations 10 (October):239-260.

Merton, Robert K.
1938 "Social structure and anomie." American Sociological Review 3 (October):672-682.

———.
1957 Social Theory and Social Structure. Glencoe, Illinois: The Free Press.

Michels, Robert.
1959 Political Parties. Trans. by Eden and Cedar Paul. New York: Dover.

Milbrath, Lester W.
1965 Political Participation. Chicago: Rand McNally.

Millen, Bruce H.
1963 The Political Role of Labor in Developing Countries. Washington, D.C.: The Brookings Institution.

Miller, S. M.
1960 "Comparative social mobility, a trend report and bibliography." Current Sociology 9 (No. 1):1-89.

Mills, C. Wright.
1956 White Collar. New York: Oxford University Press.
Minucci, Adalberto, and Saverio Vertone.
1960 Il Grattacielo nel Deserto. Roma: Editori Riuniti.
Mizruchi, Harold Ephraim.
1960 "Social structure and anomie in a small city." American Sociological Review 25 (October):645-654.
Moore, Wilbert E.
1951 Industrialization and Labor. Ithaca, New York: Cornell University Press.
———.
1963 "Industrialization and social change." Pp. 299-370 in Bert F. Hoselitz and Wilbert E. Moore (eds.), Industrialization and Society. The Hague: UNESCO-Mouton.
———.
1965 The Impact of Industry. Englewood Cliffs, New Jersey: Prentice-Hall, Inc.
———.
1966 "Changes in occupational structure." Pp. 194-212 in Neil J. Smelser and Seymour Martin Lipset (eds.), Social Structure and Mobility in Economic Development. Chicago: Aldine Press.
———.
1967 Order and Change: Essays in Comparative Sociology. New York: John Wiley and Sons, Inc.
Moore, Wilbert E., and Arnold S. Feldman (eds.).
1960 Labor Commitment and Social Change in Developing Areas. New York: Social Science Research Council.
Morris, David Morris.
1960 "Labor market in India." Pp. 173-200 in Wilbert E. Moore and Arnold S. Feldman (eds.), Labor Commitment and Social Change in Developing Areas. New York: Social Science Research Council.
———.
1968 "Labor relations: Developing countries." Pp. 510-516 in David L. Sills (ed.), International Encyclopedia of the Social Sciences. New York: Macmillan and The Free Press.
Morse, Nancy C., and Robert S. Weiss.
1955 "The function and meaning of work and the job." American Sociological Review 20 (April):191-198.

Murphey, Rhodes.
1969 "Urbanization in Asia." Pp. 58-75 in Gerald Breeze (ed.), The City in Newly Developing Countries. Englewood Cliffs, New Jersey: Prentice-Hall, Inc.

Myers, Charles A.
1958 Labor Problems in the Industrialization of India. Cambridge, Massachusetts: Harvard University Press.

———.
1959 "India." Pp. 19-74 in Walter Galenson (ed.), Labor and Economic Development. New York: Wiley.

Myers, Charles A., and Subbiah Kannappan.
1970 Industrial Relations in India. Bombay: Asia Publishing House.

Nakane, Gchie.
1970 Human Relations in a Vertical Society: Theory of a Unitary Society. Berkeley and Los Angeles: University of California Press.

Nash, Manning.
1958 Machine Age Maya: The Industrialization of a Guatemalan Community. Glencoe, Illinois: The Free Press.

Nettler, Gwynn.
1965 "A further comment on 'anomy.'" American Sociological Review 30 (October):762-763.

Neufeld, Maurice F.
1954 Labor Unions and National Politics in Italian Industrial Plants. Ithaca, New York: Cornell University Press.

Niehoff, Arthur.
1959 Factory Workers in India. Milwaukee, Wisconsin: Milwaukee Public Museum Publications in Anthropology, No. 5.

Niemeyer, Glenn A.
1963 The Automotive Career of Ransom E. Olds. East Lansing, Michigan: Michigan State University, Bureau of Business and Economic Research, Graduate School of Business Administration.

Nosow, Sigmund.
1956 "Labor distribution and the normative order." Social Forces 35 (October):25-33.

Okochi, Kazuo, Bernard Karsh, and Solomon B. Levine (eds.).
1974 Workers and Employers in Japan. Princeton: Princeton University Press.

Olsen, Marvin E.
1969 "Two categories of political alienation." Social Forces 47 (March):288-299.

Ornati, Oscar.
1955 Jobs and Workers in India. Ithaca, New York: Cornell University Press.

———.
1963 "The Italian economic miracle and organized labor." Social Research 30 (Winter):519-526.

Palmieri, Horacio, and Rinaldo Antonio Colome.
1964 "La industria manafacturera en la ciudad de Córdoba." Revista de Economia y Estadistica (Nos. 3 and 4):35-71.

Parker, Richard.
1972 The Myth of the Middle Class. New York: Harper and Row.

Pendle, George.
1963 Argentina. London: Oxford University Press.

Portes, Alejandro.
1971 "Political primitivism, differential socialization, and lower-class leftist radicalism." American Sociological Review 36 (October):820-835.

Portes, Alejandro, and Adreain Ross.
1974 "A model for the prediction of leftist radicalism." Journal of Political and Military Sociology 2 (Spring):33-56.

Przeworski, Adam, and Henry Teune.
1970 The Logic of Comparative Social Inquiry. New York: Wiley-Interscience.

Rajagopalan, C.
1962 The Greater Bombay: A Study in Suburban Ecology. Bombay: Popular Book Depot.

Reder, Melvin W.
1968 "Wages: Structure." Pp. 403-414 in David L. Sills (ed.), International Encyclopedia of the Social Sciences. New York: Macmillan and The Free Press. Vol. 16.

Report of Special Task Force to the Secretary of Health, Education, and Welfare.
1973 Work in America. Cambridge, Massachusetts: MIT Press.

Reynolds, Lloyd G., and Peter Gregory.
1965 Wages, Productivity, and Industrialization in Puerto Rico. Homewood, Illinois: R. D. Irwin.

Roberts, Alan H., and Milton Rokeach.
1956 "Anomie, authoritarianism, and prejudice: A replication." American Journal of Sociology 61 (January):355-358.

Roethlisberger, F. J., and William J. Dickson.
1947 Management and the Worker. Cambridge, Massachusetts: Harvard University Press.

Rose, Arnold M.
1959 Indagine Sull'Integrazione Social in Due Quartieri di Roma. Roma: Istituto di Statistica, Università di Roma.

Russett, Bruce M., et al.
1964 World Handbook of Political and Social Indicators. New Haven, Connecticut: Yale University Press.

Sayles, Leonard R.
1963 Behavior of Industrial Work Groups: Prediction and Control. New York: John Wiley and Sons.

Scobie, James R.
1964 Argentina: A City and a Nation. New York: Oxford University Press.

Scoville, James G.
1974 "Afghan labor markets: A model of interdependence." Industrial Relations 13 (October):274-287.

Seeman, Melvin.
1959 "On the meaning of alienation." American Sociological Review 24 (December):783-791.

———.
1971 "The urban alienations: Some dubious theses from Marx to Marcuse." Journal of Personality and Social Psychology 19 (August):135-143.

Sewell, William H., Archibald O. Haller, and Alejandro Portes.
1969 "The educational and early occupational attainment process." American Sociological Review 34 (February):82-92.

Sewell, William H., Archibald O. Haller, and George W. Ohlendorf.
1970 "The educational and early occupational status attainment process: Replication and revision." American Sociological Review 35 (December):1014-1027.

Sharma, Baldev R.
1969 "The Indian industrial workers." International Journal of Comparative Sociology 10 (March-June):161-177.

———.
1974 The Indian Industrial Worker. Delhi: Vikas Publishing House.

Shepard, Jon M.
1971 Automation and Alienation. Cambridge, Massachusetts: MIT Press.

Sheppard, Harold L., and Neil Herrick.
1972 Where Have All the Robots Gone? New York: The Free Press.

Sheth, N. R.
1968 The Social Framework of an Indian Factory. Manchester: Manchester University Press.

Shostak, Arthur B.
1969 Blue-Collar Life. New York: Random House.

Signorini, Mario.
1970 "Aspetti e fasi dell 'integrazione del contadino nella società urban." Rassegna Italiana di Sociologìa 11 (January-March): 121-141.

Simmel, Georg.
1955 Conflict and the Web of Group Affiliations. Trans. by Kurt H. Wolff and Reinhard Bendix. Glencoe, Illinois: The Free Press.

Simpson, Miles E.
1970 "Social mobility, normlessness and powerlessness in two cultural contexts." American Sociological Review 35 (December):1002-1013.

Smelser, Neil J.
1959 Social Change in the Industrial Revolution. Chicago: University of Chicago Press.

———.
1963a "Mechanisms of change and adjustment to change." Pp. 33-54 in Bert F. Hoselitz and Wilbert E. Moore (eds.), Industrialization and Society. New York: UNESCO-Mouton.

———.
1963b The Sociology of Economic Life. Englewood Cliffs, New Jersey: Prentice-Hall.

Smelser, Neil J., and Seymour Martin Lipset.
1966 "Social structure, mobility and development." Pp. 1-50 in Neil J. Smelser and Seymour Martin Lipset (eds.), Social Structure and Mobility in Economic Development. Chicago: Aldine Publishing Company.

Smith, Joel, William H. Form, and Gregory P. Stone.
1954 "Local intimacy in a middle-sized city." American Journal of Sociology 60 (November):276-284.

Smith, Michael A.
1968 "Process technology and powerlessness." British Journal of Sociology 19 (March):76-88.

Soares, Glaucio A. D.
1966 "Economic development and class structure." Pp. 190-197 in Reinhard Bendix and Seymour Martin Lipset (eds.), Class, Status, and Power. New York: The Free Press.

Sovani, N. V.
1966 Urbanization and Urban India. Bombay: Asia Publishing House.

Spenner, Kenneth I.
1975 "The internal stratification of the working class: A reanalysis." American Sociological Review 40 (August):513-520.

Spinrad, William.
1960 "Correlates of trade union participation: A summary of the literature." American Sociological Review 25 (April):237-244.

Srinivas, M. N.
1969 Social Change in Modern India. Berkeley and Los Angeles: University of California Press.

Srole, Leo.
1956 "Social integration and certain correlates: An exploratory study." American Sociological Review 21 (December):709-716.

Stinchcombe, Arthur L.
1961 "Agricultural enterprise and rural class relations." American Journal of Sociology 67 (September):165-176.

Stone, Gregory P.
1954 "City shoppers and urban identification." American Journal of Sociology 60 (July):36-45.

Straus, Murray A.
1969 "Phenomenal identity and conceptual equivalence of measurement in cross-national comparative research." Journal of Marriage and the Family 31(May):233-239.

Sussman, Marvin B.
1959 "The isolated nuclear family: Fact or fiction." Social Problems 6 (Spring):333-340.

Taira, Koji
1970 Economic Development and the Labor Market in Japan. New York: Columbia University.

Tannenbaum, Arnold S., and Robert L. Kahn.
1958 Participation in Union Locals. Evanston, Illinois: Row, Peterson.

Tausky, Curt.
1969 "Meanings of work among blue collar workers." Pacific Sociological Review 12 (Spring):49-55.

Taviss, Irene.
1969 "Change in the form of alienation: The 1900's vs. the 1950's." American Sociological Review 34 (February):46-57.

Thomlinson, Ralph.
1969 Urban Structure: The Social and Spatial Character of Cities. New York: Random House.

Tilgher, Adriano.
1930 Work: What It Has Meant to Men Through the Ages. Trans. by Dorothy Canfield Fisher. New York: Harcourt, Brace.

Toennies, Ferdinand.
1957 Community and Society. Trans. and edited by Charles P. Loomis. East Lansing, Michigan: Michigan State University Press.

Touraine, Alain.
1955 L'Evolution du Travail Ouvrier aux Usines Renault. Paris: Centre National de la Recherche Scientifique.

Touraine, Alain, and Orietta Ragazzi.
1961 Ouvriers d'Origine Agricole. Paris: Aux Editions du Seuil.

Tudor, Bill.
1972 "A specification of relationships between job complexity and powerlessness." American Sociological Review 37 (October): 596-604.

Udy, Stanley H., Jr.
1971 Work in Traditional and Modern Society. Englewood Cliffs, New Jersey: Prentice-Hall.

United Nations.
1961 Statistical Yearbook.

U.S. Bureau of the Census.
1960 Women by Number of Children Ever Born. PC(2)3A. Washington, D.C.: U.S. Government Printing Office, U.S. Department of Commerce.

———.
1962 "Detailed characteristics." Final Report, PC(1)-24D. Washington, D.C.: U.S. Government Printing Office.

———.
1966. "Michigan area statistics." Pp. 23-28 in Census of Manufactures: 1963. Washington, D.C.: U.S. Government Printing Office.

———.
1971 Pocket Data Book, U.S.A., 1971. Washington, D.C.: U.S. Government Printing Office.

Vaidyanathan, K. E.
1971 "Work force in Greater Bombay: Social and demographic characteristics." (Unpublished manuscript.)

Vroom, V. H.
1964 Work and Motivation. New York: Wiley.

Walker, Charles R., and Robert Guest.
1952 The Man on the Assembly Line. Cambridge, Massachusetts: Harvard University Press.

Ware, Norman J.
1935 Labor in Modern Industrial Society. Boston: D. C. Heath.

Westley, William A., and Margaret W. Westley.
1971 The Emerging Worker. Montreal: McGill-Queen's University Press.

Whyte, William F.
1944 "Who goes union and why." Personnel Journal 23 (December):215-230.

Whyte, William Foote, et al.
1955 Money and Motivation. New York: Harpers.

Wiebe, Paul.
1973 "Annanagar—social life in a Madras slum." (Unpublished manuscript.)

Wilensky, Harold L.
1956 Intellectuals in Labor Unions. Glencoe, Illinois: Free Press.

———.
1961a "Life cycles, work situation and participation in formal associations." Pp. 213-242 in Robert W. Kleemeier (ed.), Aging and Leisure. New York: Oxford University Press.

———.
1961b "Orderly careers and social participation: The impact of work history on social integration in the middle mass." American Sociological Review 26 (August):521-539.

———.
1962 "Labor and leisure: Intellectual traditions." Industrial Relations 1 (February):1-12.

Wilensky, Harold L., and Hugh Edwards.
1959 "The skidder: Ideological adjustments of downward mobile workers." American Sociological Review 24 (April):215-231.
Windmuller, John P.
1974 "European labor and politics: A symposium (1)." Industrial and Labor Relations Review 28 (October):3-6.
———.
1975 "European labor and politics: A symposium (11)." Industrial and Labor Relations Review 28 (January):203-207.
Woodward, Joan.
1958 Management and Technology. London: H.M.S.O.
Wright, Charles R., and Herbert Hyman.
1958 "Voluntary association memberships of American adults: Evidence from national sample surveys." American Sociological Review 23 (June):284-294.
Wyatt, S., and R. Marriott.
1967 A Study of Attitudes to Factory Work. London: H.M.S.O.
Zeitlin, Maurice.
1967 Revolutionary Politics and the Cuban Working Class. Princeton, New Jersey: Princeton University Press.
Zimmer, Basil G.
1955 "Participation of migrants in urban structure." American Sociological Review 20 (April):218-224.
Zwerman, William L.
1970 New Perspectives on Organization Theory. Westport, Connecticut: Greenwood Publishing Co.

Author Index

Subject Index

LIBRARY OF CONGRESS CATALOGING IN PUBLICATION DATA

Form, William Humbert, 1917-
Blue-collar stratification.

Bibliography: p.
Includes index.
1. Automobile industry workers. 2. Industrial sociology.
3. Machinery and industry. I. Title.
HD8039.A8F67 301.44′42 75-17425
ISBN 0-691-09366-0